Water Resources and Environmental Engineering I

Maheswaran Rathinasamy · S. Chandramouli
K. B. V. N. Phanindra · Uma Mahesh
Editors

Water Resources and Environmental Engineering I

Surface and Groundwater

Springer

Editors
Maheswaran Rathinasamy
Department of Civil Engineering
Maharaj Vijayaram Gajapathi
Raj College of Engineering
Vizianagaram, Andhra Pradesh, India

S. Chandramouli
Department of Civil Engineering
Maharaj Vijayaram Gajapathi
Raj College of Engineering
Vizianagaram, Andhra Pradesh, India

K. B. V. N. Phanindra
Department of Civil Engineering
Indian Institute of Technology Hyderabad
Hyderabad, Telangana, India

Uma Mahesh
Department of Civil Engineering
National Institute of Technology Warangal
Warangal, Telangana, India

ISBN 978-981-13-2043-9 ISBN 978-981-13-2044-6 (eBook)
https://doi.org/10.1007/978-981-13-2044-6

Library of Congress Control Number: 2018951405

This Springer imprint is published by the registered company Springer Nature Singapore Pte Ltd.
The registered company address is: 152 Beach Road, #21-01/04 Gateway East, Singapore 189721, Singapore

Foreword

Judicious management of water resources is fundamental for achieving sustainable management of natural resources and ensuring environmental integrity. Technologies, such as remote sensing, navigation, space communication, geospatial tools, Internet of things, are extremely useful in developing newer applications and tools for scientific data management and decision making.

The international conference organized by the Department of Civil Engineering, MVGR College of Engineering (A), Vizianagaram, from 30 March to 01 April 2018 provided a much-needed platform to discuss the emerging technologies and opportunities in water, environment and climate change facets.

The effort of the organizers in bringing out a scientific book on conference deliberations and a compendium of papers needs a special compliment.

I strongly believe that the technical insights presented in this book will enrich the scientific community and provide inspiration to readers and lead to newer technological applications that would support human society in coping up with the challenges posed by impending climate change.

I wish the organizing committee of the conference a grand success.

Hyderabad, India

Y. V. Krishna Murthy
Director
National Remote Sensing Centre

Preface

With the ever-increasing demand for development, the stress on water resources and environment is increasing day by day. The changing climate further amplifies the effect resulting in severe drought, flood and pollution problems. In order to provide a platform for eminent scientists, researchers and students to discuss the emerging technologies in mitigating the problems related to water and environment, the International Conference on Emerging Trends in Water Resources and Environmental Engineering (ETWREE 17) was conducted by MVGR College of Engineering, Vizianagaram, Andhra Pradesh, India, during Mar–Apr 2017. About 100 participants from three different countries attended ETWREE 17. ETWREE 17 was organized by the Department of Civil Engineering, MVGR College of Engineering, and was sponsored by Science and Engineering Research Board (SERB) and National Remote Sensing Centre (NRSC).

The proceedings of this conference contain 60 papers which are included as two volumes. The response to ETWREE 17 was overwhelming. It attracted quality work from different areas relating to water resources, environmental engineering and climate. From a total of 120 abstracts, we selected around 80 papers through a rigorous peer review process with the help of our programme committee members and external reviewers for the presentation.

Dr. Y. V. N. K. Murthy, Director, NRSC Hyderabad, conducted a special session on "Application of Remote Sensing in Water Resources". A special session on "Enigma of Climate" was conducted by Prof. Rakesh Khosa, IIT Delhi. Professor D. Nagesh Kumar from IISC Bangalore delivered a lecture on "Remote Sensing, GIS and DEM for Water Resources Assessment of a River Basin". Professor Uma Mahesh, NIT Warangal, gave a lecture on "Non-Stationarity in Rainfall Intensity". Dr. Brijesh Kumar Yadav, IIT Roorkee, conducted a session on "Engineered BioRemediation". Dr. K. B. V. N. Phanindra, IIT Hyderabad, delivered a keynote on "Modeling Soil Water Disease Interactions of Flood Irrigated Mandarin Orange Trees".

Dr. Shishir Gaur, IIT BHU, conducted a special session on "Application of Simulation Optimization Model for Management of Groundwater Resources". Dr. L. Suri Naidu, NUS Singapore, delivered a lecture on "Food, Water and Energy

Nexus". Professor G. V. R. Srinivas Rao, Andhra University, conducted a session on "Multivariate Statistical Analysis of River Water Quality". Professor T. V. Praveen, Andhra University, delivered a lecture on "Salinity Intrusion Modelling". Dr. Y. R. S. Rao, NIH Kakinada, provided a lecture on "River Bank Filtration".

These sessions were very informative and beneficial to the authors and delegates of the conference. We thank all the keynote speakers and the session chairs for their excellent support to make ETWREE 17 a grand success. The quality of a contributed volume is solely due to the reviewers' efforts and dedication. We thank all the members of the advisory board of the conference for their support and encouragement.

We are indebted to the programme committee members, Mr. A. V. S. Kalyan, Mr. Varaprasad and Mr. Sridhara Naidu, for extending their help in preparing the manuscript.

We express our heartfelt thanks to the Chief Patron, Sri Ashok Gajapathi Raju, Chairman, MANSAS, and Patron, Prof. K. V. L. Raju, Principal, MVGR College of Engineering, for their continuous support and encouragement during the course of the convention. We also thank all the faculty and administrative staff for their efforts.

We would also like to thank the authors and participants of this conference, who have made it for the conference. Finally, we would like to thank all the student volunteers who spent their assiduous efforts in meeting the deadlines and arranging every detail to make sure the smooth running of the conference. All the efforts are worth if the readers of this contributed volume find them inspiring and useful. We also sincerely thank the press, print and electronic media for their excellent coverage of this convention.

Vizianagaram, India — Dr. Maheswaran Rathinasamy
Vizianagaram, India — Dr. S. Chandramouli
Hyderabad, India — Dr. K. B. V. N. Phanindra
Warangal, India — Prof. Uma Mahesh
December 2017

About This Book

The book covers a variety of topics related to water, climate and environment. The topics mainly focus but not limited to hydrological modelling, water resources management, water conservation practices, applications of recent techniques for solving water-related issues, land use impact on water resources, climate change impacts, wastewater treatment and recovery, advances in hydraulics in rivers and ocean. The book is a collection of best papers submitted in the First International Conference on Emerging Trends in Water Resources and Environmental Engineering held from 28 March 2017 to 1 April 2017 at MVGR College of Engineering, Vizianagaram, Andhra Pradesh, India. It was hosted by the Department of Civil Engineering, MVGR College of Engineering, with the support of Science and Engineering Research Board, India.

Contents

About the Editors

Dr. Maheswaran Rathinasamy is currently Associate Professor, Department of Civil Engineering, MVGR College of Engineering, Vizianagaram. He received his bachelor's and master's degree from Anna University, Chennai, and BIT Mesra, respectively. He obtained his Ph.D. from IIT Delhi. He is a recipient of INSPIRE Fellowship from the Department of Science and Technology, India, and Humboldt Fellowship from Alexander Von Humboldt Foundation, Germany. He has postdoctoral experience in the University of Minnesota, USA, and Potsdam Institute of Climate Impact Research, Germany. He is principal investigator of funded research projects on the order of 1.5 crore rupees. He has around 30 international journal publications and 25 international conference publications. His research interests include stochastic hydrology, hydrological modelling and hydro-meteorological forecasting.

Dr. S. Chandramouli currently serves as Professor and HOD, Department of Civil Engineering, MVGR College of Engineering, Vizianagaram. He received his M.Tech. with water resources engineering as specialization from NIT Warangal in 2002. He obtained his Ph.D. in civil engineering from Andhra University, Visakhapatnam, in 2013. He has worked in several organizations such as CES(I) Pvt. Ltd., Hyderabad; GVP College of Engineering, Visakhapatnam; and GMRIT, Rajam, for a period of 10 years. He is working with MVGR College of Engineering since 2011. He has published more than 50 technical papers in various reputed journals and conferences. He has attended more than 60 professional training programmes organized by prestigious institutions in India. He is the life member of ISTE and IEI. He has completed one DST project as a co-principal investigator. He has reviewed many journal papers published by prestigious journals and conferences. He has organized many faculty development programmes and student training programmes.

Dr. K. B. V. N. Phanindra currently serves as Assistant Professor of Civil Engineering at IIT Hyderabad, India. He received his master's degree in hydraulics and water resources engineering from IIT Kanpur and Ph.D. in water resources engineering from New Mexico State University (NMSU). He also holds a graduate minor degree in GIS from NMSU. To his credit, he has nine journal publications of international repute, three technical reports, one monograph and one chapter. He has completed three research projects funded by various ministries from the Government of India to the tune of about 1.6 crore rupees. His research interests include hydrogeologic characterization, groundwater flow and transport modelling, soil–water–crop interactions, remote sensing and GIS applications in groundwater.

Prof. Uma Mahesh is currently serving as Professor in the Department of Civil Engineering at National Institute of Technology, Warangal, Telangana, India. He has earlier served as Head of the Department from July 2008 to June 2010, as Dean, Students' Welfare from July 2012 to March 2013 and as Dean, Planning & Development from April 2013 to June 2014. His area of specialization is water resources with a focus on water resources systems, hydrologic modelling, irrigation management, water quality modelling and management, applications of soft computing techniques and modelling impacts of climate change. He is a recipient of the Jalamitra Award by the Government of Andhra Pradesh in 2003 for successful implementation of Watershed Development Project in Warangal District, G. M. Nawathe Award for the paper presented at Hydro 2004 (annual conference of the Indian Society for Hydraulics) and Central Board of Irrigation and Power (CBIP) Award. Eight Ph.D. students have graduated with Prof. Uma Mahesh as their advisor. He is currently advising six Ph.D. students at NIT Warangal. He has published more than 60 papers in various reputed journals and conferences.

What Constitutes a Fair and Equitable Water Apportionment?

Himanshu Tyagi, A. K. Gosain and Rakesh Khosa

Abstract Water has been a source of conflict since time immemorial. Numerous mechanisms have been proposed for solving such conflicts but multiplicity of water uses and users along with self-serving definition of equitable, makes dispute resolution challenging. Doctrines advocating water appropriation based on the notion of equity and fairness are intuitively appealing. However, subjectivity of this concept impedes their translation to universal principles for water allocation as fairness quotient of any mechanism is determined unitedly by gamut of diverse factors. Thus, the present study critically reviews the connotations of equity and equality to arrive at a procedurally and distributionally just apportionment policy for real-world water conflicts. It seeks an equal opportunity paradigm for deservedness-based resource distribution that could be unanimously amenable to all stakeholders. The study is very apposite as there is a lurking fear of heightened water conflicts that could have bitter socio-political ramifications.

Keywords Conflict resolution · Egalitarianism · Equity and fairness
Proportionality · Transboundary rivers

1 Introduction

Water is undoubtedly one of the most indispensable resources for sustaining life on this planet. With 40% of global population residing within 263 international transboundary river basins in 145 countries [1], there have always been tensions over sharing water resources. Moreover, with burgeoning demands for freshwater and deteriorating sources of supply, there is a lurking fear that there will be a rise in occurrence and intensity of such water conflicts [2].

Transboundary water disputes are of multi-disciplinary nature and involve an array of natural, hydrological, social, political, and economic issues [3, 4].

H. Tyagi (✉) · A. K. Gosain · R. Khosa
Department of Civil Engineering, Indian Institute of Technology Delhi, New Delhi, India
e-mail: tyagiben@gmail.com

M. Rathinasamy et al. (eds.), *Water Resources and Environmental Engineering I*,
https://doi.org/10.1007/978-981-13-2044-6_1

Gamut of factors like geography, hydrological spatio-temporal variability, population pressure, unsustainable utilization, vested interests, geo-politics, industrialization, budding human expectations, etc., can play a pivotal role in triggering a possible conflict in any communal interstate or international basin [5–7].

Most transboundary rivers are shared between just two countries, but there are about 13 basins which have 5–8 stakeholder nations. While rivers like Congo, Niger, Nile, Rhine and Zambezi have 9–11 riparians, Danube River navigates through 18 countries [1]. With such multiplicity of stakeholders, chances of arriving at consensus diminish [8]. It has also been observed that the contenders deliberately overvalue their strong attributes and underrate their negative characteristics to get advantageous outcomes [9]. Further, as everybody considers themselves to be more rational [10, 11], dispute resolution becomes challenging due to vested interests and justice bias [12]. Nevertheless, if conflicts remain unresolved, eventually there may be trust deficit issues which might do severe long-lasting socio-political damages. However, it may be noted that so far no war has been fought over water as it is neither hydrographically effective nor economically worthwhile [8].

In the history of transboundary conflicts, several approaches have been employed to resolve differences over water sharing through negotiations, public consultations, third party adjudication/arbitration, decree, water markets, river basin authorities, decision support systems, etc. [2, 13]. Also, various water laws and doctrines have evolved from historical practices of handling shared resources. However, subjective understanding of such principles by researchers and administrators has proved to be a serious impediment in formulation of universal water allocation doctrines.

Naturally, doctrines advocating appropriations based on the paradigms of equity and fairness, carry great appeal. But objective translation of this concept beyond philosophy has not been very successful [14] and consequently it is difficult to decide entitlements in real world conflicts on the basis of this vague concept. Therefore, this study intends to define the concepts of equity and fairness for an equitable allocation policy that can be used to resolve water sharing conflicts.

2 Theory of Equity and Fairness

The concept of equity and fairness is a worldwide social concern and is therefore intuitively appealing since time immemorial. Traditionally, the idea of equity was limited to the professions of law, public welfare and social sciences [7]. But its connotation and application changed gradually with evolving socio-political scenarios, and is now pertinent for administrators, economists and scientists too who often grapple to interpret equity in their respective fields. For example, administrators essentially look for equity in affirmative actions [15] and employment plans [16]. In economics, equity is a key issue in studies involving distribution of income [17]. Researchers exploring the idea of water rights principally study the notions of equity [14].

Studies from diverse areas have often used the terms, equity and fairness, interchangeably considering their similar scope and definition. While Webster describes equity as fairness, impartiality, justice; Oxford English Dictionary defines fair as equitably, honestly, impartially, justly, according to rule.

The twin concept of equity and fairness is vague and idiosyncratic as every individual has his own perceptions of equity and fairness, and consequently there is barely any consensus on its precise and objective articulation. Literature review reveals that researchers from different disciplines have struggled to develop an objective definition of equity. Marsh and Schilling [18] presented a framework to choose the most suitable measure of equity from the existing equity measures. But there has been a continuous argument on whether a method is equitable or not as multiple issues and parameters determine the equity quotient of any proposed distribution mechanism [19].

Peyton Young in his book, Equity: In Theory and Practice [17], states that the notion of equity is multifaceted and thus cannot be easily defined. He says that to define equity for a particular case, contextual details must be considered too as equity is greatly influenced by stakeholder attributes, social beliefs, precedents and the resource being distributed. He was of the opinion that equity helps in determining the most appropriate outcome based on uniformity and neutrality, and thus it legitimates the allocation choice. He suggests to consider following questions before arriving at any equity solution:

- What form should the allocation take?
- What are the eligibility criteria?
- What counts in the distribution?
- What are the relevant principles?
- What are the relevant precedents?
- How should competing principles and criteria be reconciled?
- What incentives does a rule create?

3 Envy and Superfairness

Superfairness analysis originates from games like fair division of cake in which one person gets the opportunity to cut the cake in two parts whereas the other person chooses the slice he wants. In this case, a distribution is termed superfair if both the persons fancy their own share more than the share received by each other, i.e., nobody envies each other [20]. Varian [21] calls such an allocation equitable while terming an equitable as well as Pareto optimal distribution as fair.

A distribution principle that allots water among various co-riparians on the basis of a particular criterion or a specific combination of criteria that confers a certain advantage only to a particular riparian is destined to create envy amongst the other claimants and hence cannot lead to a consensual solution. Tinbergen [22] proposed the idea of an envy-free equitable system in which nobody wants to be in somebody else's position. But this concept was not realistic as envy is an inherent trait in humans due to which they always compare themselves with others and then try to compensate their weaknesses with any of their positive attributes. Foley [23] suggested a more practical approach stating that it is not essential to have an envy-free society but nobody should prefer anybody else's allocation. For instance, if different fruits are being distributed, everyone should prefer his own fruit over the fruits that others got on belief that they got the fruit that they desired the most.

4 Normative Theory of Justice and Aristotle's Maxim

Moulin [24] defined distributive fairness through Nicomachean Ethics based Aristotle's famous adage: 'Equals should be treated equally and unequals unequally, in proportion to relevant similarities and differences'. However, Bazerman et al. [25] highlighted the difficulties associated with definition and measurement of equity suggested by Moulin [24].

Equal treatment of equals' advocates that if the claimants have same characteristics in all the relevant areas, they should get the same share in the resource being distributed. In contrast, the principle of unequal treatment is ambiguous but it can be said that it suggests that the resources should be shared in a proportion that highlights the differences between the claimants or in other words, the deservedness of the stakeholders [26].

5 Procedural and Distributive Justice

Any fair distribution mechanism should address the following two concerns, namely, (i) Is the distribution fair? (ii) Is the outcome fair? While the first question relates to procedural justice, the second question examines the distributive justice [24].

Psychologists studied resource sharing from the perspective of exchange between different individuals and these studies led to the development of procedural justice concepts. Furthermore, the aspiration for equity in social justice schemes resulted in the theory of distributive justice. Apropos social welfare policy, Rasinski [27] reported that equity has two elements, viz., proportionality and egalitarianism. While the former recommends individual apportionments based on individual deservedness, the latter involves equal opportunities in resource distribution.

According to equality principle, everyone should be treated equally. However, it may be noted that equality does not necessarily entail equal allocation for all stakeholders but is more suggestive of the distribution process involved. Thus, equality ensures procedural justice and it can be said that if the process is just and equal, then the resulting allocations are likely to be easily accepted by the stakeholders [28]. Proportionality is an established norm to ensure distributive justice

[17]. The proportionality doctrine advocates resource allocation based on claimant's contribution to that resource measured on a cardinal scale.

In series of studies, the authors Syme, Nancarrow and McCreddin [29–31] developed socio-psychological theories of justice, equity and fairness for water allocation decision-making and presented the correlation between procedural and distributive justice. These studies emphasized on the importance of environmental, economic and social issues for attaining sustainability. Instead of social impact assessments of different water policies, the studies evaluated the fairness of different outcomes as an indicator for social criteria.

Syme and Nancarrow [29] assessed the ethical considerations that are relevant to water allocation systems. They conducted primary investigation in three areas: (i) philosophical basis for deciding allocations, (ii) attitude towards planning approaches, and (iii) concept of procedural justice.

Using major equity, distributive justice and procedural justice variables identified through previous studies, Syme and Nancarrow [30] also examined the apparent fairness of water allocation decision-making through a questionnaire survey administered on water literate people. The study showed that people can take assertive decisions on fairness after getting well-versed with procedural and distributive aspects of the system under examination. Also, following observations were made: (i) water seen as a public good, (ii) environment seen to have water rights, (iii) procedural issues are important in water allocation decision-making, and (iv) situational fairness is also important.

Syme et al. [31] conducted studies to find those measures which can reveal how people evaluate justice, equity and fairness. The authors concluded that the participants' notions are likely to change with time and hence the fairness heuristic may vary temporally according to the socio-political dynamics.

6 Conclusion

The preceding discussion may be summarized as follows:

- Subjective nature of water sharing principles severely hinders the formulation of universal water apportionment doctrines.
- The concept of equity and fairness is nebulous and can hardly be objectively articulated.
- Equity does not essentially mean equal distribution of the sought resource, but it implies an equal opportunity paradigm where everyone is equal before the law and the actual distribution is based on deservedness.
- Procedural justice requires an equal opportunity decision-making process that is perceived as open, transparent and unprejudiced.
- Distributional justice involves equitable resource allocation. Proportionality or deservedness is often seen as a rational model for distribution of a given resource.

- Water apportionment purely on the basis of a particular criterion may generate envy among co-riparians and therefore a holistic approach is required to fix the quantum of allocations.

Based on above conclusions, it can be recommended that a coercion-free platform should be given to all the stakeholders of a water dispute wherein each of them can propose a quantifiable criterion that can be most advantageous to him in deservedness-based entitlements. A multi-criteria mathematical formulation should then be used to derive respective proportional apportionments of each claimant. This decision-making mechanism is not only egalitarian but is distributively just also, and thus can be effectively used to resolve transboundary water allocation conflicts.

References

1. Water, U.N.: Transboundary Waters: Sharing Benefits, Sharing Responsibilities. UN Water, Zaragoza, Spain (2008)
2. Singh, A.: Conflict Resolution in Transboundary Watercourses through Integration of an Operationalised Legal Doctrine with GIS Based Hydrological Modelling. Doctoral thesis, Department of Civil Engineering, IIT-Delhi, New Delhi (2008)
3. Le-Huu, H.: Potential water conflicts and sustainable management of international water resources systems. Water Resour. J. 1–13 (2001)
4. Tamas, P.: Water Resource Scarcity and Conflict: Review of Applicable Indicators and Systems of Reference, vol. 1. UNESCO (2003)
5. Lazerwitz, D.J.: The flow of international water law: the international law commission's law of the non-navigational uses of international watercourses. Indiana J. Global Legal Stud. 247–271 (1993)
6. Biswas, A.K.: Management of shared natural resources: problems and prospects. J. Indian Water Resour. Soc. **3**(1), 7–18 (1983)
7. Ravikumar, K.: Towards a Fair and Feasible Allocation of Cauvery Waters. Doctoral thesis, Department of Civil Engineering, IIT-Delhi, New Delhi, India (2007)
8. Wolf, A.T.: Conflict and cooperation along international waterways. Water Policy **1**(2), 251–265 (1998)
9. Tyler, T., Hastie, R.: The social consequences of cognitive illusions. Res. Negot. Organ. **3**, 69–98 (1991)
10. Messick, D.M., Bloom, S., Boldizar, J.P., Samuelson, C.D.: Why we are fairer than others. J. Exp. Soc. Psychol. **21**(5), 480–500 (1985)
11. Taylor, S.E., Brown, J.D.: Illusion and well-being: a social psychological perspective on mental health. Psychol. Bull. **103**(2), 193 (1988)
12. Cohen, R.L.: Justice and negotiation. Res. Negot. Organ. **3**, 259–282 (1991)
13. Tyagi, H., Gosain, A.K., Khosa, R.: Mechanisms for resolving transboundary water conflicts. In: Virtual Poster Showcase Abstracts (2017)
14. Sampath, R.K.: Equity measures for irrigation performance evaluation. Water Int. **13**(1), 25–32 (1988)
15. Nacoste, R.W.: The effects of affirmative action on minority persons: research in the Lewinian tradition. In: The Lewin Legacy, pp. 268–281. Springer New York (1986)
16. Pfeffer, J., Langton, N.: Wage inequality and the organization of work: the case of academic departments. Adm. Sci. Q. 588–606 (1988)
17. Young, H.P.: Equity: In Theory and Practice. Princeton University Press (1995)

18. Marsh, M.T., Schilling, D.A.: Equity measurement in facility location analysis: a review and framework. Eur. J. Oper. Res. **74**(1), 1–17 (1994)
19. Gleick, P.H.: Water and conflict: fresh water resources and international security. Int. Secur. **18**(1), 79–112 (1993)
20. Baumol, W.J.: Super Fairness: Applications and Theory. MIT Press, Massachusetts (1987)
21. Varian, H.R.: Equity, envy, and efficiency. J. Econ. Theor. **9**(1), 63–91 (1974)
22. Tinbergen, J.: Redeljke Inkomensverdeling. N.V. DeGulden Pers, Haarlem (1953)
23. Foley, D.K.: Resource allocation and the public sector. Yale Econ. Essays **7**, 45–98 (1967)
24. Moulin, H.: Fair Division and Collective Welfare. MIT Press, Cambridge, MA (2003)
25. Bazerman, M.H., Wade-Benzoni, K.A., Tenbrunsel, A.E.: Psychological difficulties in resolving water disputes. In: UNESCO International Workshop on Negotiations Over Water: Conflicts, Results, Techniques (1998)
26. Tyagi, H., Gosain, A.K., Khosa, R., Anand, J.: Paradigm for distributive & procedural justice in equitable apportionment of transboundary Ganges waters under changing climate & landuse. In: AGU Fall Meeting Abstracts (2015)
27. Rasinski, K.A.: What's fair is fair—or is it? value differences underlying public views about social justice. J. Pers. Soc. Psychol. **53**(1), 201 (1987)
28. Pandey, D.P.: Towards a Fair and Equitable Allocation of Krishna Waters. M.Tech thesis, Department of Civil Engineering, IIT-Delhi, New Delhi (2015)
29. Syme, G.J., Nancarrow, B.E.: Planning attitudes, lay philosophies, and water allocation: a preliminary analysis and research agenda. Water Resour. Res. **32**(6), 1843–1850 (1996)
30. Syme, G.J., Nancarrow, B.E.: The determinants of perceptions of fairness in the allocation of water to multiple uses. Water Resour. Res. **33**(9), 2143–2152 (1997)
31. Syme, G.J., Nancarrow, B.E., McCreddin, J.A.: Defining the components of fairness in the allocation of water to environmental and human uses. J. Environ. Manage. **57**(1), 51–70 (1999)

Impact of Anthropogenic Interventions on the Vembanad Lake System

Raktim Haldar, Rakesh Khosa and A. K. Gosain

Abstract Estuarine and coastal zone processes have always been topic of research due to their being prime centers of rich resources like diverse habitat and natural beauty. Other than ecological reasons these aquatic bodies act as important economic centers, tourist places, serve in navigational purposes, and fishing. One of the India's most valued natural sites is the Vembanad Lake and estuarine system that lies on the western coast in the state of Kerala. This natural system, which comprises the lake, the Kuttanad wetland region and the Cochin estuary, is included in the Ramsar list of important wetland sites. Six major rivers, namely, Periyar, Muvattupuzha, Pamba, Manimala, Meenachil, and Achenkovil contribute to the system. The whole system has been vastly modified throughout the last couple of centuries owing to sedimentation and human-driven factors. On the other hand, there has been constant reclamation of the low-lying areas on the periphery of the lake and the wetlands, leading to reduction in the spread area. The special characteristics of these lands that lie to the east of the lake is that the ground level is lower than the lake water level. Therefore, the lake water easily serves for irrigational purpose in these adjacent lands. According to tentative proposals in the recent years it was intended to make further developments in the catchment areas for various purposes. The present paper takes a modeling approach to find out what would be the possible impact on the lake water profile as well as salinity/solute concentration if these proposals are implemented. The study has been carried out using the two-dimensional hydrodynamic modeling software MIKE 21 with HD and AD modules. The results from the hydrodynamic model of the lake, although not fully representative, show that the lake water levels and salinity might face quantitatively relevant changes which can pose a threat to the natural environment.

Keywords Wetland · Lake · Estuary · Modeling · Ecosystem
Hydrodynamics · Vembanad

R. Haldar (✉) · R. Khosa · A. K. Gosain
Indian Institute of Technology Delhi, New Delhi, India
e-mail: rhaldar.iitd@gmail.com

M. Rathinasamy et al. (eds.), *Water Resources and Environmental Engineering I*,
https://doi.org/10.1007/978-981-13-2044-6_2

1 Introduction

Water being a limited resource, and having users in multiple sectors such as agricultural, domestic, and industrial, the competitive interaction between its supply and demand prevails in all places of the world. Especially in India, where the population has risen at a high rate, demand for more water and at times, occurrence of floods, have forced us to create interventions that would somehow make the situation in hand more suitable in accordance with our needs. With the passage of time we have created many such changes which are irreversible. It is recently that we have started to understand the ecological aspect of water resources [1], as compared to history of agricultural and industrial development. Governments in developing countries like India have recently started to consider environmental flows and wetland conservation in framing water laws and regulations. Undoubtedly, we should attempt to understand the natural systems to the best of our ability and then take further steps towards development or modification in them. The Vembanad Lake (Fig. 1) in the state of Kerala is one such aquatic system which has suffered vast amount of anthropogenic interventions in the past couple of centuries [2].

The environmental condition of the Vembanad Lake has been in steady decline due to various anthropogenic activities. The major problems are (i) decrease in water holding capacity of the lake; (ii) weed infestation (iii) decrease in water

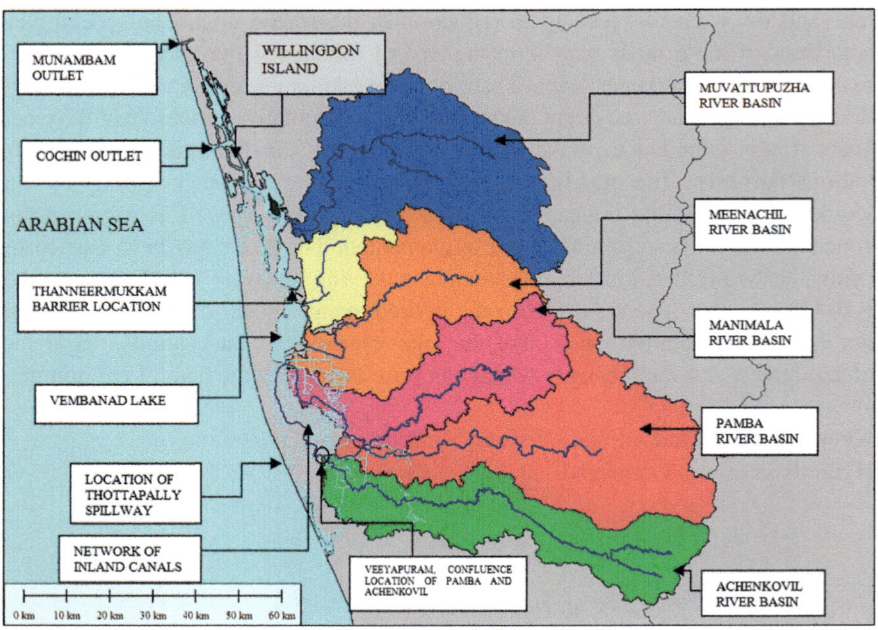

Fig. 1 The Vembanad Lake along with the contributing river basins *Muvattupuzha, Meenachil, Manimala, Pamba* and *Achenkovil*

quality through both organic and inorganic loading; and (iv) shrinkage of habitat. Pouring of industrial effluents, retting practice, houseboats, etc., can be indicated as few of the contributors to pollution in the lake. The lake has seen decades of wetland reclamation, and construction of structural interventions for controlling floods and saline intrusion, greatly changing its physical characteristics, producing a resultant threat to the ecosystem [3]. Most of the issues faced by the lake and adjacent wetland system, which have been recognized, have been addressed by [4] and [5]. Major research has been done in estimation of the lake water quality and its fish count [6–10]. Studies like [11–14] account harmful impacts of anthropogenic activities on the flora and fauna of aquatic ecosystems in other parts of the world.

Being a coastal estuarine system, and fed by fresh water from rivers, the saline characteristics of Vembanad lake and estuary system naturally follows a typically annual cyclic pattern. Several studies have investigated the salinity pattern in the estuary and its variation along the estuary [15–18]. In general, the lake salinity pattern varies between three phases in a year, both in terms of salinity level and stratification. During the low-flow season in the rivers (February–May), the salinity values are high; with the advent of rainy season, as the freshwater contribution from the rivers increases, the lake becomes relatively less saline (June–September); and as the freshwater contribution reduces (October–January) the salinity rises gradually [19].

In a coastal wetland system such as Vembanad, various factors like currents, tides, waves, turbulence, light, temperature, salinity, bed materials and suspended particles determine the transport and dispersion of all suspended and dissolved material in a lake, including contaminants and nutrients, and affect all biological activity from the transport of larvae and bacteria and the growth of phytoplankton to the behavior of fishes. Though every intricate activity and their associated complex relations are difficult to capture, hydrodynamic modeling, to some extent, provides the required platform for ecological investigations. Prediction of flow properties connected to such phenomena has been one of the growing interests in the last few decades in the scientific community supported by the growing computational capabilities of the generation. Previous attempt for hydrodynamic modeling of the Cochin Estuary has been made by [20–23], etc.

The current research is an attempt to estimate the impact of two previously proposed man-made interventions on the flow and salinity characteristics of the lake, through modeling. The first intervention considered is the Pamba-Achenkovil-Vaippar Link Project and the second is Thottapally channel expansion plan. The first intervention was planned as a part of the National inter-basin river-linking plan developed by National Water Development Agency (NWDA). The second intervention was planned under the Kuttanad Development Package, aimed at expanding the carrying capacity of the channel leading to Thottapally spillway through deepening and/or widening [24]. The existing channel which was constructed in 1955 is responsible for draining of water from rivers Pamba, Achenkovil, and Manimala to the sea. It was constructed to reduce the magnitude of floods in the Kuttanad so as to allow year-round agricultural practice in the area [25]. However, the constructed channel was inefficient in its carrying capacity and therefore it was proposed to increase its carrying capacity [26]. In this study, the MIKE 21 hydrodynamic (HD) and advection-dispersion (AD) tools have been used to derive the flow characteristics of the lake for the existing system and to estimate the past and future scenario using simulation. The study is somewhat limited by paucity of available data.

2 Study Area

The Vembanad lake plus wetland system lies in the Indian state of Kerala, between latitudes 9°5′ N and 10° N and longitudes 76°15′ E and 76°45′ E. Spread across the districts of Ernakulam, Kottayam, and Alappuzha, the lake consists of a complex system of backwaters, marshes, lagoons and an intricate network of canals along with the main water body, henceforth endowing Alappuzha the name "Venice of the East". Other than providing water for paddy cultivation, the lake serves other important purposes like aquaculture, tourism, and transport. The Vembanad system can be divided into three parts owing to its morphology, dynamics and man-made structural interventions. The northern estuary part extends from Azhikode/Munambam to north of Cochin outlet. The southern estuary ranges from Cochin inlet to Thanneermukkom where a barrage was made in 1975 to prevent salt intrusion. The southernmost main lake body remains fresh throughout most of the year, protected by the Thanneermukkom Barrage (TB). At the Cochin outlet, the channel is 450 m wide and 8–12 m deep and connects the Vembanad system to the Arabian Sea on the western coast. Six major rivers provide input to the lake, namely, Periyar, Muvattupuzha, Meenachil, Manimala, Pamba, and Achenkovil. While the Periyar River has upstream interventions diverting a part of its water towards the neighboring state Tamil Nadu, other rivers like Manimala, Pamba, and Achenkovil have downstream channelization which diverts a part of their water to the sea during floods. The other part of the Vembanad system, from Cochin to Munambam, is contributed mainly by Periyar River and drains into the Sea mainly through Munambam Beach outlet. The Muvattupuzha River joins the system in the middle estuary part while the other four join in the southern part. These five rivers originate from the Southwestern Ghat ranges. The annual rainfall in the region is above 3000 mm. The various soil types are Coastal Alluvium, Forest loams, Lateritic soil, Brown hydromorphic, Riverine Alluvium, Greyish Onattukara, etc., (CGWB district reports, December 2013).

2.1 The Thottapally Spillway

Before the extensive reclamation of the wetland, the Kuttanad region used to be flooded almost every year during the monsoons due to the high discharges in Pamba and accompanying rivers. However, in order to mitigate the floods and hence protect the paddy cultivation, the Thottapally Spillway was built in 1955, which was expected to divert excess floodwaters brought by Pamba directly to the Arabian Sea, without traversing the distance up to Vembanad Lake.

2.2 *The Thanneermukkom Salt Water Barrier*

The Vembanad Lake and estuary used to be saline in the non-monsoon period of the year. During the rainy season, in the months of July to September, the lake surface is inhibited by freshwater from the rivers [19, 27, 28]. The seasonal variation of salinity in the lake allowed only one-time paddy cultivation in the year. Therefore, in 1976, the Thanneermukkom salt water barrier was constructed in order to control the salinity of the Vembanad Lake, in order to intensity paddy cultivation and hence make better use of the reclaimed wetland area. The Thanneermukkom barrier or bund runs a total length of about 1400 m. It can be visualized as composed of three parts, where the middle one-third part is a coffer dam, and the rest two-third (both sides) consists of shutters.

Figure 2 shows the location and details of Thanneermukkom Bund. The shutters of the barrage/bund are operated so as to maintain the salinity in the Vembanad

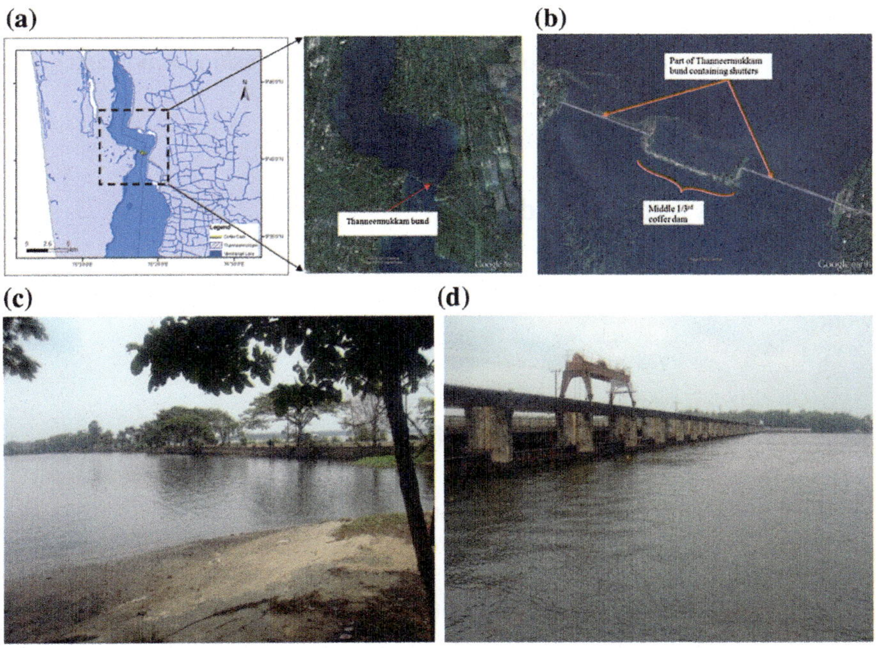

Fig. 2 **a** Location of Thanneermukkom Bund in the Vembanad Lake; **b** Image of Thanneermukkom bund from Google Earth (*Source* Google Earth.); **c** A photograph showing 'Coffer dam' part of Thanneermukkom bund (*Date* March 10, 2012); **d** A photograph showing part Thanneermukkom bund containing shutters (*Date* March 10, 2012)

Table 1 Thanneermukkom operation schedule for 10 years 2001–2011

Sl. No.	Year	Closing	Opening
1	2001–02	1-Nov-2001	4-Feb-2002
2	2002–03	23-Dec-2002	24-Mar-2003
3	2003–04	17-Dec-2003	20-Mar-2004
4	2004–05	12-Dec-2004	17-Mar-2005
5	2005–06	23-Dec-2005	25-Mar-2006
6	2006–07	22-Dec-2006	4-May-2007
7	2007–08	15-Dec-2007	4-Jul-2008
8	2008–09	23-Dec-2008	31-Mar-2009
9	2009–10	21-Dec-2009	26-Mar-2010
10	2010–11	1-Mar-2010	5-Mar-2011

Lake (south of Thanneermukkom) below 2 ppt (parts per thousand), in order to facilitate unimpaired growth of paddy. The barrier (shutters) remains closed during the low-flow season. The dates of shutter closure and opening in the last 10 years are given in Table 1. Although the barrage resulted in reducing the salinity of lake water south of Thanneermukkom, its establishment proved somewhat harmful for aquaculture [9, 29, 30]. It was found by [31], who compared the salinity profiles of the estuary before and after the commissioning of the barrier, that salinity has markedly decreased on the southern side of the Thanneermukkom bund and the water body remains fresh for most of the year. However the southern part of the lake now remains polluted due to the ceasing of the tidal flushing action and growth of weeds has been observed [4].

3 Materials and Methods

This section consists of the tools and methodology that have been employed in the work.

3.1 MIKE 21 Hydrodynamic Model

The hydrodynamic model simulates unsteady two-dimensional flows in one layer (vertically homogeneous) fluids. It solves the vertically integrated equations for conservation of mass and momentum.

The model domain was discretized with grid size 180 m $\times$ 180 m covering lake area 250 km^2. The lake bathymetry was developed from the river navigational sheets obtained from Inland Waterways Authority of India (IWAI) plus surveyed data made available by WRD, GoK (Water Resources Dept., Govt. of Kerala) (Fig. 3). Maximum depth in the model domain was around 12 m at Cochin outlet,

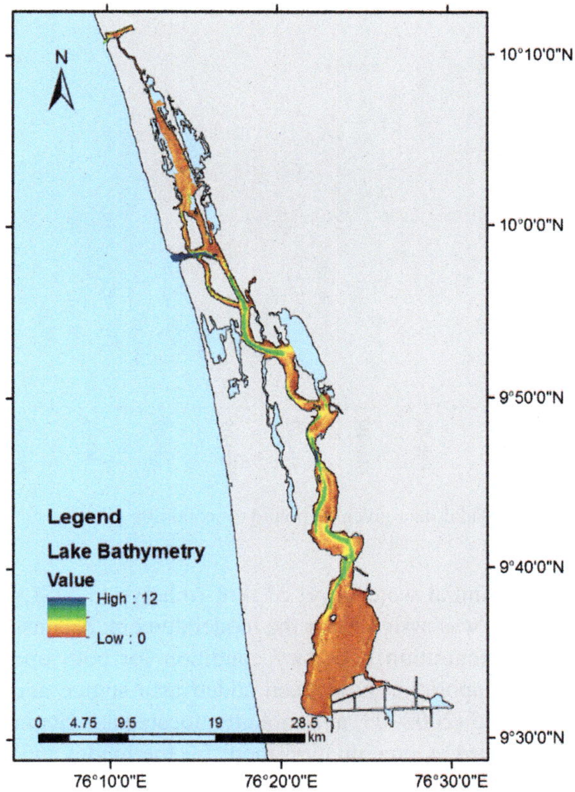

Fig. 3 Vembanad Lake and estuary Bathymetry

whereas the lowest depths are found near the Kumarakom area, southernmost in the domain.

Tidal data was obtained from Cochin Port Trust with the help of WRD, GoK only for the Cochin outlet location (Fig. 4). This was used as elevation type boundary condition for the HD model. Based on a parametric study, a suitable value of Manning's number was found to be $M = 25\ \mathrm{m}^{1/3}/\mathrm{s}$, which was applied over the whole model domain. The simulation time step was 5 s for model stability. The value of dispersion coefficient varies across different water bodies and also spatio-temporally in the same water body and is a property difficult to measure. For rivers, longitudinal and transversal dispersion coefficient values have been suggested to be ranging from 7.6×10^{-1} to $1.5 \times 10^{3}\ \mathrm{m}^2/\mathrm{s}$ and from 4.8×10^{-3} to $1.1\ \mathrm{m}^2/\mathrm{s}$, respectively [32]. For lakes, dispersion coefficient values have been suggested to range from 10^{-3} to $10^{-2}\ \mathrm{m}^2/\mathrm{s}$ [33]. However, values for coastal waters can be higher (order of magnitude $10^2\ \mathrm{m}^2/\mathrm{s}$) than the values measured in rivers due to tidal action [34]. For estuaries, values have been measured ranging from 6 to $1500\ \mathrm{m}^2/\mathrm{s}$ [32]. In the present study, a uniform value of $20\ \mathrm{m}^2/\mathrm{s}$ was assigned to the dispersion coefficient after having trials with different values.

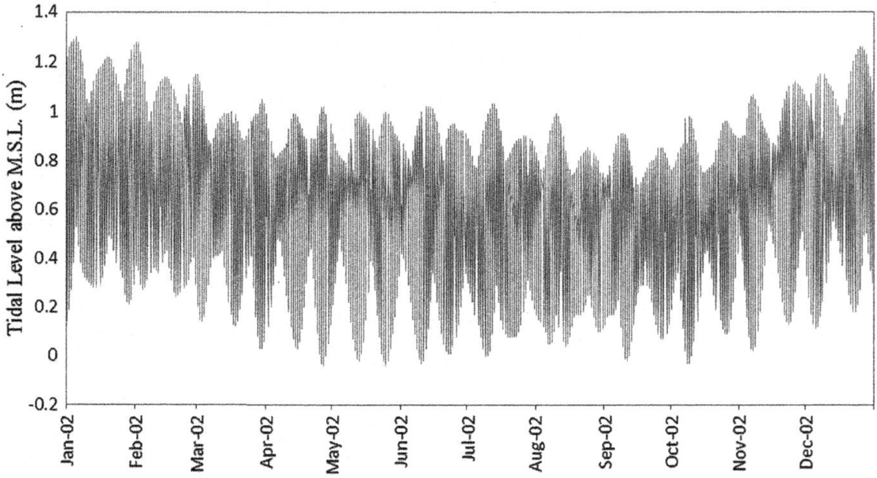

Fig. 4 Tidal data given as boundary condition

An initial water level of 0 m (relative to MSL) and an initial concentration of 1.5 ppt was assigned for the model domain. A constant value of 35 ppt was adopted as concentration boundary condition for both openings. Direct precipitation and lake evaporation have been added as a source and sink respectively. The rainfall recorded (2002–03) at Arookutty, located at a site in close proximity to the lake, has been used as precipitation input for the model and is shown plotted in Fig. 5. The temperature in Kerala normally ranges from 28 to 32 °C (82–90 °F). Loss of water

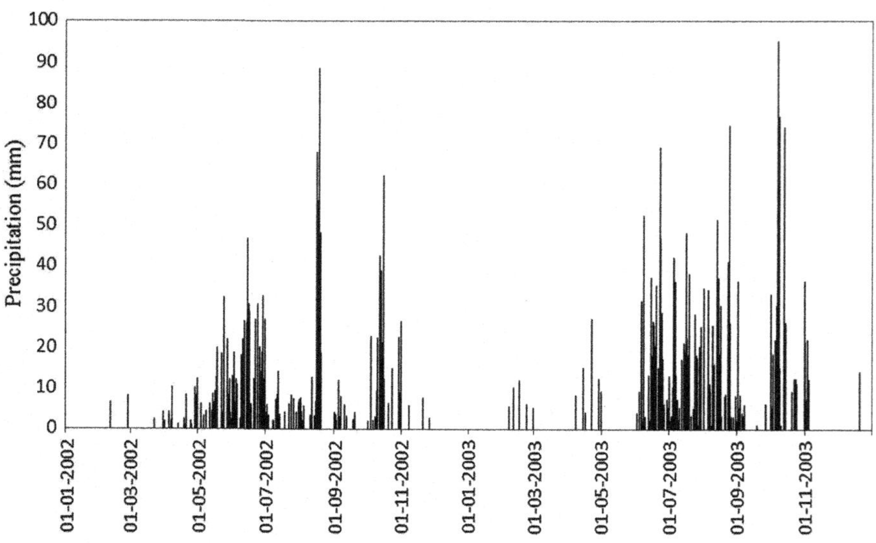

Fig. 5 *Arookutty* rainfall series for model input

from the lake due to evaporation amounts to significant proportions considering the large areal extent of the lake. Figure 6 shows the plot of the time series of the pan evaporation values that has been used as representative values for the period of hydrodynamic simulation (2002–03).

The daily flow data of the five rivers and their pollution content are obtained from ISW, Govt. of Kerala. Hydrological models for the five river basins, incorporating the impact of proposed human interventions, have been developed using SWAT model [35, 36] by [37]. The results from the above work are used for the current study for scenarios of proposed development (Figs. 7, 8 and 9).

Data pertaining to water of the five rivers was made available through the office of ISW, Govt. of Kerala and the water quality attributes for which test evaluations were available include electrical conductivity, turbidity, chloride, etc. Salinity levels of river water were obtained using the Electrical Conductivity (EC) data provided for the year of 2008–09 using: Salinity (ppt) = 0.64 × EC (micro-mhos/cm). Accordingly, the salinity levels in five rivers (acting as sources of salinity in the lake) were estimated and are tabulated in Table 2. The abnormally high value of electrical conductivity in Meenachil and Muvattupuzha in March 2009 has been attributed to sudden flash floods at the sampling station during the time of sampling. The salinity values in these two periods were approximated to the value observed for the period of May and June 2008. However, there is limitation of data for the lake area. There is no availability of water level, velocity or quality in the model domain.

The lake receives pollution from various sources such as industries, domestic, and agricultural non-point sources [4, 38–42]. According to [43] 0.104 M m^3 d^{-1} of industrial discharges and 260 t d^{-1} of domestic/organic wastes are released into the Cochin backwaters. As established by various studies these polluting substances are loaded onto the estuary mainly in the northern part of the system [44, 45].

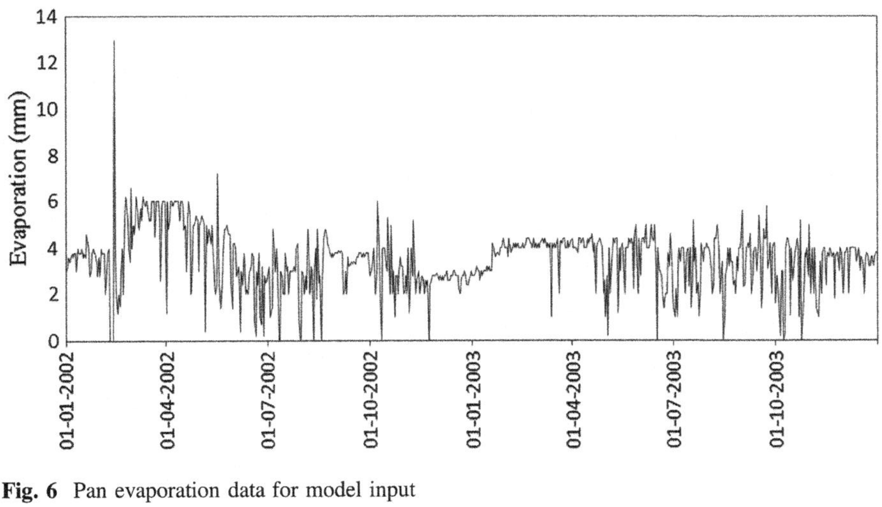

Fig. 6 Pan evaporation data for model input

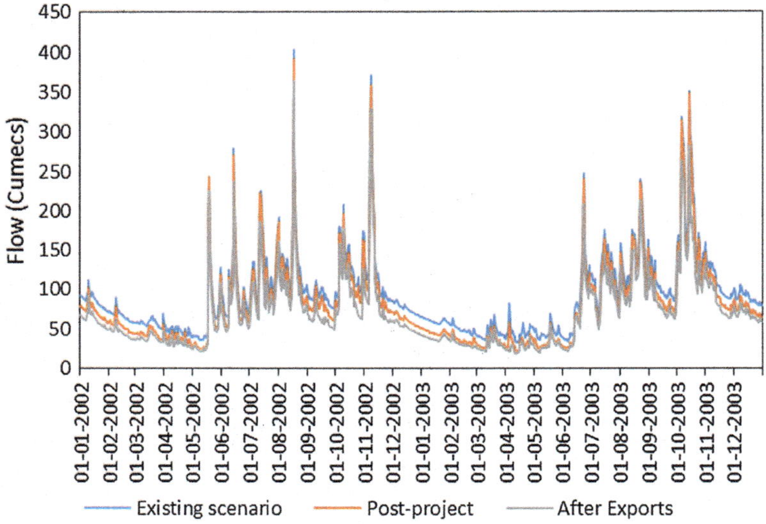

Fig. 7 Simulated daily flow for three scenarios for *Pamba* basin

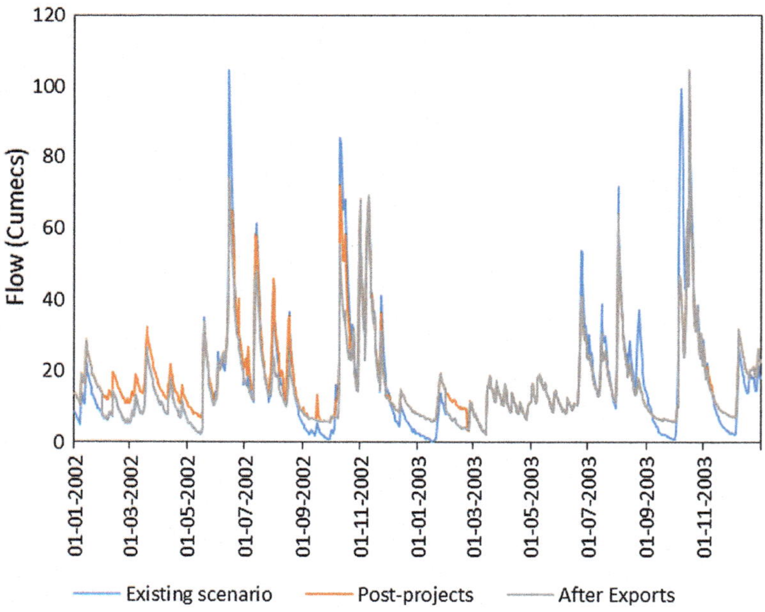

Fig. 8 Simulated daily flow for three scenarios for *Achenkovil* basin

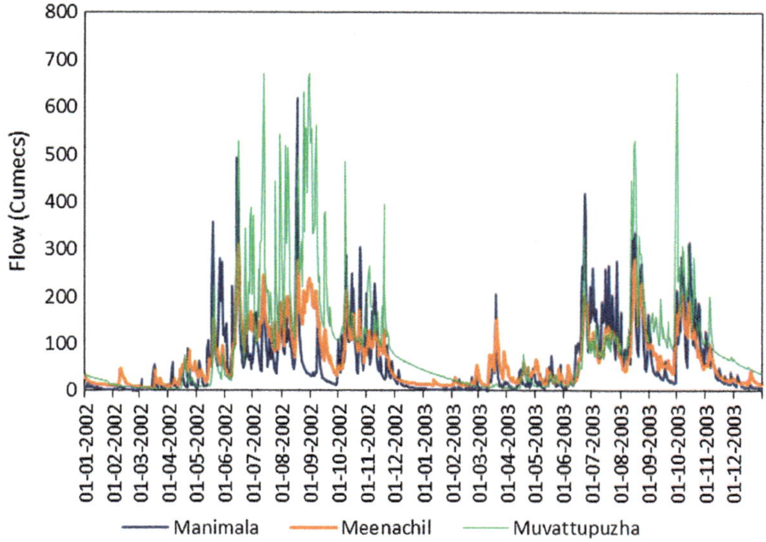

Fig. 9 Simulated daily flow for existing scenario for three river basin *Manimala*, *Meenachil* and *Muvattupuzha*

Table 2 Measured electrical conductivity values and calculated corresponding salinity values for five rivers

River	Location	Time of collection of sample	Electrical conductivity (micro-mhos/cm)	Salinity or solute concentration (ppm or mg/l)
Meenachil	Kumarakom	March 2009	2.64×10^6	1.68×10^6
		Oct and Nov 2008	51.3	32.832
		May and Jun 2008	23.7	15.168
Manimala	Kavalam	Mar 2009	61	39.04
		Oct and Nov 2008	57	36.48
		May and Jun 2008	76	48.64
Muvattupuzha	Vaikom	Mar 2009	5.96×10^6	3.81×10^6
		Oct and Nov 2008	610	390.4
		May and Jun 2008	43.9	28.096
Achenkovil	Venmoney	Mar 2009	64.4	41.216
		Oct and Nov 2008	49.3	31.552
		May and Jun 2008	56.2	35.968
Pamba	Erapuzha	Mar 2009	26.6	17.024
		Oct and Nov 2008	29.3	18.752
		May and Jun 2008	32.8	20.992

The model setup involves input of assumed effluent pour points each with a discharge of 2.5 cumecs and concentration 2.5 ppt which are evenly located at interval of 10 grid cells, southwards to northwards, along the estuary part of the system. The locations of entrance of rivers as well the assumed effluent discharge points (in form of solute concentration) have been shown in Figs. 10 and 11. Time series data is neither available for lake water level nor lake salinity.

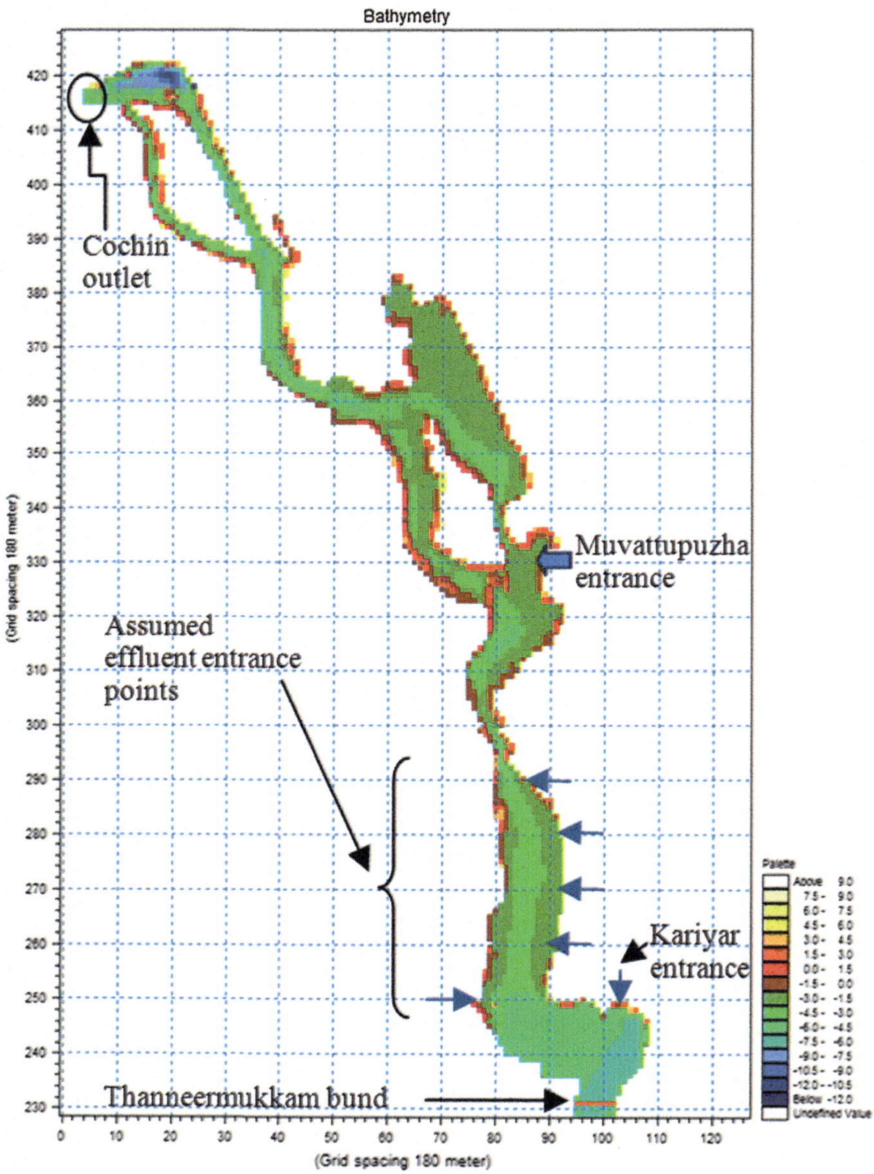

Fig. 10 Point source locations for river/drain entrance north of Thanneermukkom bund

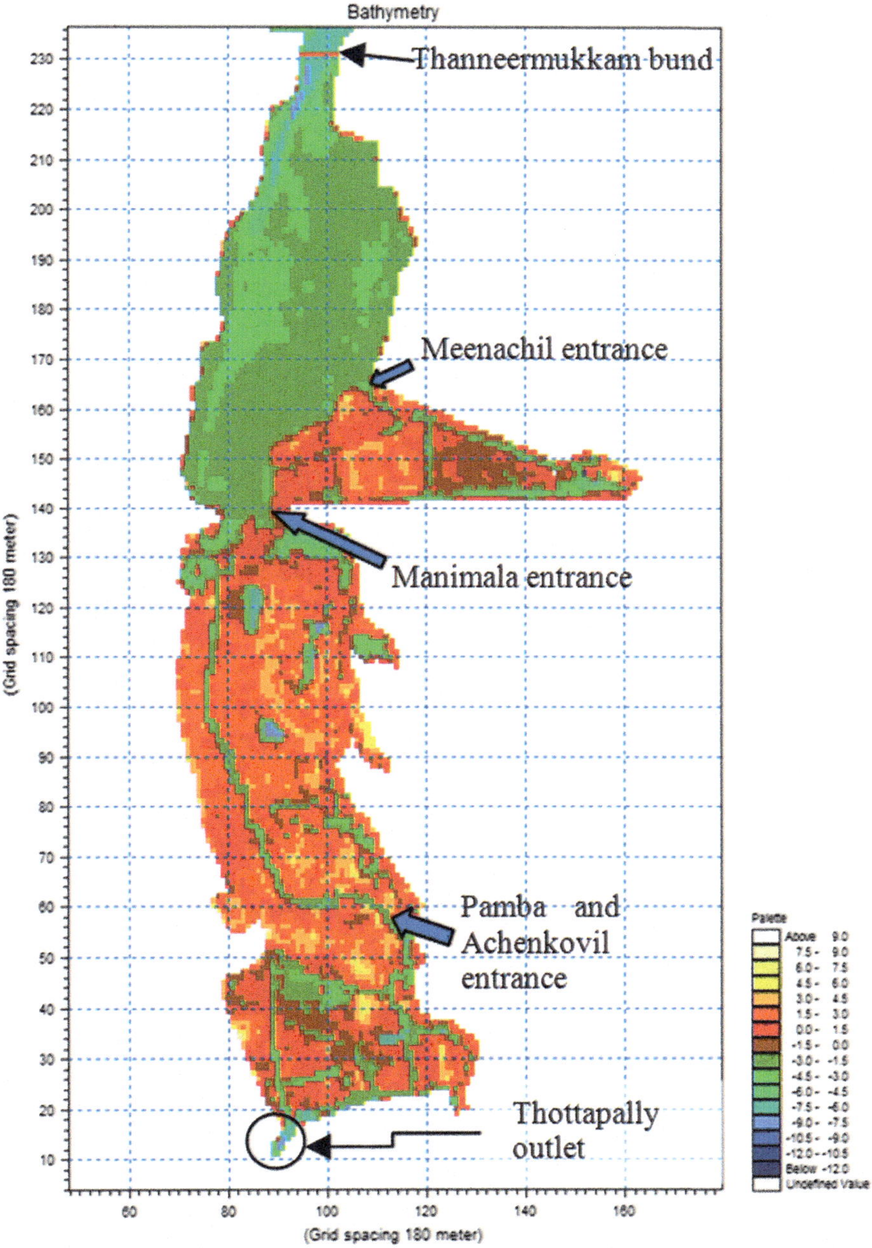

Fig. 11 Point source locations for river/drain entrance south of Thanneermukkom bund

3.2 Calibration for Existing Scenario and Simulation for the Second Scenario

In order to capture the annual recurrence pattern in the lake water solute concentration the model is run for a period of one year. Since no water level and solute concentration time series is available for calibrating the model, direct calibration is not possible. However, an indirect measure of the water quality is available from the closure and opening dates of the Thanneermukkom barrage, given the fact that the gates are closed when the solute concentration crosses above the mark of 2 ppt and opened when the same crosses below the 2 ppt mark with the advent of the high-flow season. Therefore, the model parameters were calibrated in order to match the dates of opening and closing of the barrage gates. The model was calibrated for the period of January 1–December 31 , 2002 and validated for the period January 1–December 31, 2003.

After calibrating this model for the first scenario which corresponds to the existing condition, the river flow inputs were changed and set to the flow series for the second scenario, which corresponds to inclusion of proposed developments, as generated by the SWAT model [37]. The SWAT model has been designed and calibrated for the 5 drainage basins.

4 Results and Discussion

The model results have been obtained for the three scenarios:

(a) The first case corresponds to the existing scenario.
(b) The second case corresponds to the inclusion of proposed irrigation, hydro-electric and diversion projects in the Pamba and Achenkovil basins.
(c) And the third case corresponds to the scenario of increased efficiency of Thottapally spillway in releasing floodwaters, along with projects contained in case (b).

4.1 Comparison of Water Levels

The first case consists of the results for existing scenario of water resources development in the basins. Figure 12 shows the flow depth simulated by the model for a location near the Thanneermukkom bund for the first simulation case. The figure clearly shows that the lake water level reaches maximum number of peaks in the months of July–September. Next, for the understanding the effect of the man-made developments, a comparison of the water levels has been shown in the form of differences of water levels for the cases (a), (b) and (c) in Fig. 13.

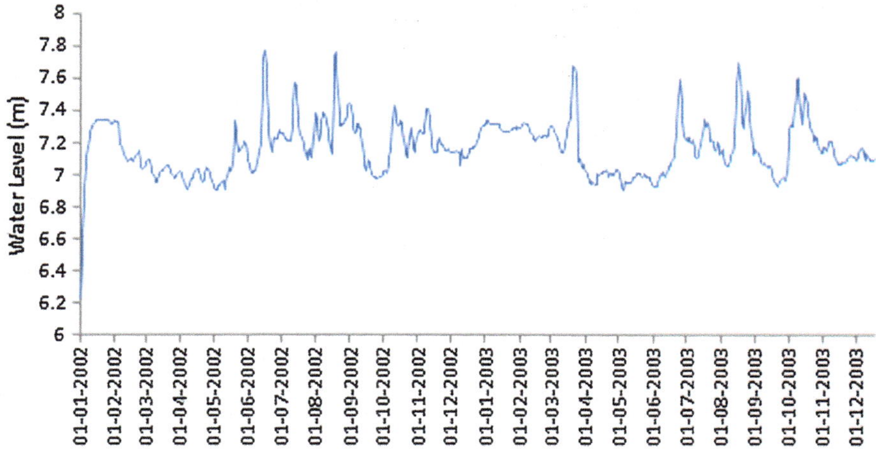

Fig. 12 Plot of simulated depth of water for a point located near Thanneermukkom bund

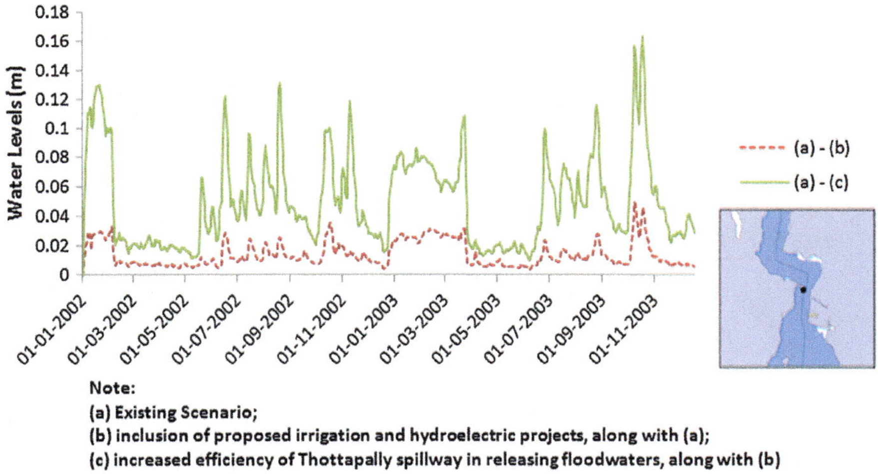

Fig. 13 Plot of reduction in water level with respect to existing scenario near Thanneermukkom bund

The maximum reduction in water level due to inclusion of projects is 0.05 m whereas that for inclusion of "increased efficiency of Thottapally spillway" case is 0.16 m.

Figures 14 and 15 show the water level changes with respect to existing scenario at two other locations. As it can be seen in Fig. 14, the plot of water level differences remain zero for a few months, that is, the change in flow volumes from the rivers Pamba and Achenkovil do not impact the water levels at this point. This is justified because the point is located north of Thanneermukkom bund, due to which

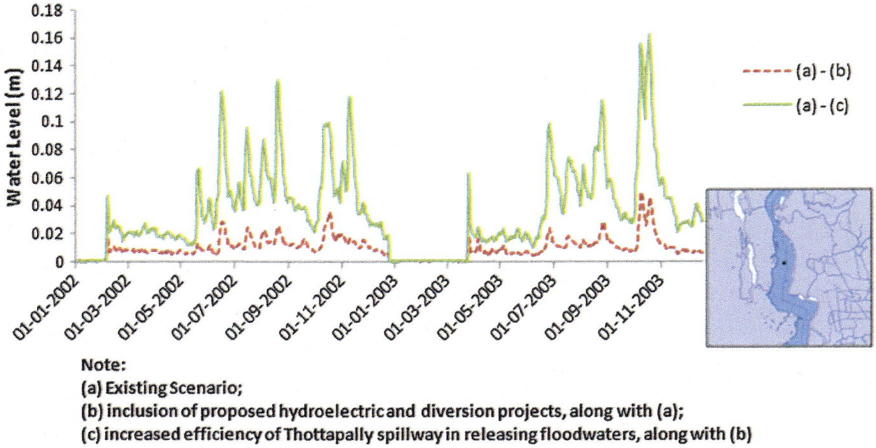

Fig. 14 Plot of change (reduction) in water level with respect to existing scenario near Vaikom

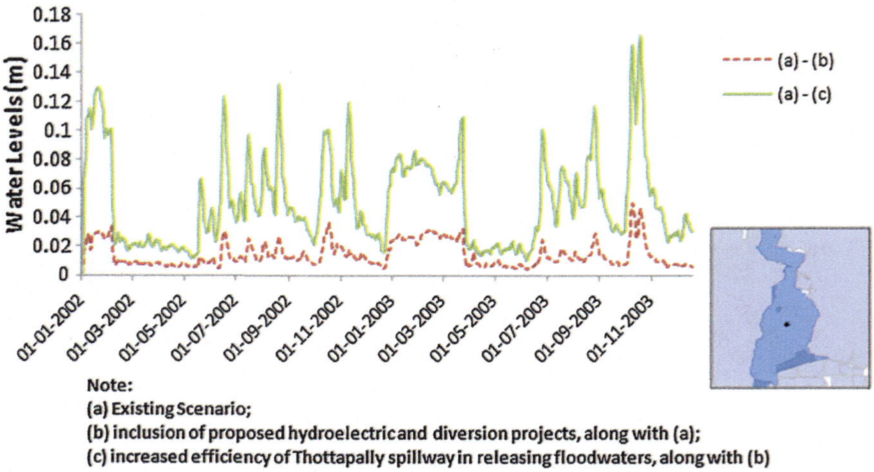

Fig. 15 Plot of change (reduction) in water level with respect to existing scenario in between the lake body

the influence of the southern river sources is negligible from December 23 to March 24, during which the bund remains closed.

The aforementioned comparisons clearly establish that the water levels in the lake are expected to register a fall of varying degrees at various locations across the lake under influence of various development scenarios and the maximum impact is expected to result if the proposal to widen the link to Thottapally spillway is carried out as envisaged. This anticipated drop in water levels has the potential to disrupt various lake processes and may result in a disturbed ecological balance in the region.

4.2 Comparison of Solute Concentration

The Vembanad Lake is surrounded by agricultural lands where paddy is cultivated. Aquaculture is practiced in the lake. There are also many varieties of industries near the estuary region. Therefore, it is a difficult task to simulate factual results due to unavailability of proper input data for pollutants. However, with the assumed pollution inputs it is possible to infer the possible influences on the water quality due to the changes in the scenario.

Figures 16 and 17 show the solute concentration results from the simulation for the three locations—near Thanneermukkom bund, and at a point in between the lake. In the two figures it is observed that

i. The solute concentration for the case (a) and (b) are more or less same.
ii. The case (c) varies widely from (a) and (b).
iii. The solute concentration differences increase during the months of January–May, which is the low-flow season.

One reason why the solute concentration corresponding to post-projects scenario (case-b) is not too different from existing scenario, even during the low-flow season, is that the Achenkovil Kal Ar reservoir and the Punnamedu reservoir which have been proposed on the Achenkovil and Pamba rivers have a recommended minimum downstream release (lean-season allowance) as per the terms of the proposal (5.72 m^3/s from Achenkovil Kal Ar reservoir and 1.43 m^3/s from Punnamedu reservoir for the period of eight months from October to May).

Because there is no measured time series of water level or solute concentration available for calibrating the model results, the model accuracy may be questioned.

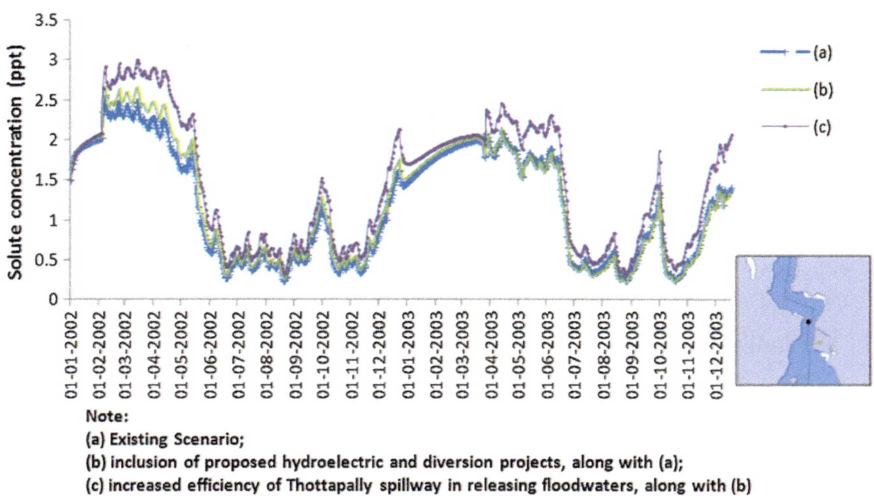

Fig. 16 Plot of simulated solute concentration level in water near Thanneermukkom bund

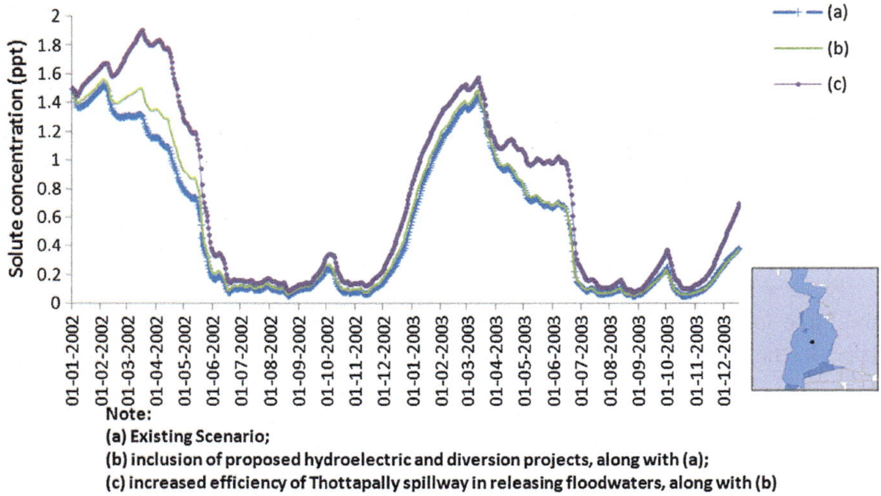

Fig. 17 Plot of simulated solute concentration level in water in between the lake body

But this exercise provides important insights pointing to the impacts of the proposed interventions.

5 Conclusion

In this study the MIKE 21 Flow Model was applied to simulate water levels and solute concentration in the Vembanad Lake. The Vembanad Lake with the Cochin estuary forms a very complex network of channels that is very difficult to model. However, it has been attempted in this study to model flow and quality aspects with various assumptions. The paper illustrates the development of a two-dimensional hydrodynamic model for the lake and estuary. The model uses the depth-integrated Navier–Stokes equations, which are called the “shallow water equations”, to find the required estimates. The model results in form of water levels and solute concentration, although are just representative because of the absence of required data for calibration, indicate that the proposed projects as well as the Thottapally lead channel expansion plan may have considerable impact on the Vembanad system.

Apart from these, the other conclusions that can be drawn from the study are

1. No further reclamation of wetland should be allowed in order to keep the ecological balance of the aquatic system safe.
2. The decision of inter-basin transfer of water should be made after accounting various possible impacts on the basin eco-hydrology. As the current study suggests there can be relevant changes in the lake and wetland characteristics as a result of the considered changes in this study. These may cause shift in the

littoral zone (due to shift in water levels) and benthic zone (due to change in the settling pattern). Further, the impact of this transfer on the saline intrusion in groundwater should also be studied.

3. Separate drainage system should be developed to intercept the wastewater coming from industrial and domestic sources, and suitable treatment plants should be set up. This will help the lake water remain cleaner throughout the year, even if the Thanneermukkom bund is closed.
4. Implementation of the proposed projects and diversions will lengthen the closure period of the Thanneermukkom barrier. As lesser freshwater will enter the lake, the solute concentration in the lake will further increase. This will affect the operation and management of both the irrigation practices and the Thanneermukkom bund.
5. In order to better capture the lake and estuary dynamics a three-dimensional model is required. Such a model will be able to represent the lake processes in a more detailed manner.

Acknowledgements The present work was performed a part of "Water Balance Study of Vembanad Wetland System", funded by Irrigation Department, Kerala. We would like to thank Mrs. P. Lathika, Chief Engineer (Inter-State Waters) and Sri Abraham Koshy (Assistant Executive Engineer), Irrigation Department, Government of Kerala for their valuable support in facilitating site visits and data collection. We thank IWAI and Cochin Port Trust for their support. We would also like to thank Aayush (M.Tech, IIT Delhi) for his contribution to this study.

References

1. Karr, J.R.: Biological integrity: a long-neglected aspect of water resource management. Ecol. Appl. **1**, 66–84 (1991)
2. Gopalan, U.K., Vengayil, D.T., Udayavarma, V.P., Krishnankutty, M.: The shrinking backwaters of Kerala. J. Mar. Biol. Assoc. India Cochin **25**, 131–141 (1983)
3. Thomson, K.T.: Economic and social issues of biodiversity loss in Cochin backwaters. In: Technical report, pp. 51–82. Cochin University of Science and Technology Cochin, India (2002)
4. Menon, N.N., Balchand, A.N., Menon, N.R.: Hydrobiology of the Cochin backwater system–a review. Hydrobiologia **430**, 149–183 (2000)
5. Sreejith, K.A.: Human impact on Kuttanad wetland ecosystem-An overview. Int. J. Sci. Technol. **2**, 670–679 (2013)
6. Asha, C.V., Suson, P.S., Retina, C.I., Nandan, S.B.: Decline in diversity and production of exploited fishery resources in Vembanad wetland system: strategies for better management and conservation. Open J. Mar. Sci. **4**, 344–357 (2014)
7. Asha, C.V, Cleetus, R.I., Suson, P.S., Nandan, S.B.: Environmental factors structuring the Fish assemblage distribution and production potential in Vembanad estuarine system, India. Int. J. Mar. Sci. **5** (2015)
8. Kurup, B.M., Sebastian, M.J., Sankaran, T.M., Rabindranath, P.: Exploited fishery resources of the Vembanad lake. Indian J. Fish. **40**, 199–206 (1993)
9. Pillai, V.K., Ponniah, A.G., Vincent, D., David Raj, I.: Acidity in Vembanad lake causes fish mortality. Mar. Fish. Inf. Serv. Tech. Ext. Ser. **53**, 8–15 (1983)

10. Vincy, M.V., Rajan, B., Kumar, A.P.P.: Water quality assessment of a tropical wetland ecosystem with special reference to Backwater Tourism, Kerala, South India. Int. J. Environ. Sci. **1**, 62–68 (2012)
11. Verschuren, D., Johnson, T.C., Kling, H.J., Edgington, D.N., Leavitt, P.R., Brown, E.T., Talbot, M.R., Hecky, R.E.: History and timing of human impact on Lake Victoria, East Africa. Proc. R. Soc. London B Biol. Sci. **269**, 289–294 (2002)
12. Otiang'a-Owiti, G.E., Oswe, I.A.: Human impact on lake ecosystems: the case of Lake Naivasha, Kenya. Afr. J. Aquat. Sci. **32**, 79–88 (2007)
13. Nouri, J., Danehkar, A., Sharifipour, R.: Evaluation of ecotourism potential in the northern coastline of the Persian Gulf. Environ. Geol. **55**, 681–686 (2008)
14. Nouri, J., Fatemi, M.R., Danekar, A., Fahimi, F.G., Karimi, D.: Determination of environmentally sensitive zones along Persian Gulf coastlines through geographic information system. J. Food, Agri. Env. **7**, 718–725 (2009)
15. George, M.J., Kartha, K.N.: Surface salinity of Cochin backwater with reference to tide. J. Mar. Biol. Assoc. India. **5**, 178–184 (1963)
16. Ramamirtham, C.P., Muthusamy, S., Khambadkar, L.R.: Estuarine oceanography of the Vembanad lake Part I: the region between Pallipuram (Vaikom) and Thevara (Cochin). Indian J. Fish. **33**, 85–94 (1986)
17. Renosh, P.R., Rasheed, K., Balchand, A.N.: Studies on tide depended salt-silt wedge and identification of turbidity maxima in Cochin estuary. Indian J. Mar. Sci. **39**, 136–142 (2010)
18. Shivaprasad, A., Vinita, J., Revichandran, C., Reny, P.D., Deepak, M.P., Muraleedharan, K R., NaveenKumar, K.R.: Seasonal stratification and property distributions in a tropical estuary (Cochin estuary, west coast, India) (2013)
19. Qasim, S.Z., Gopinathan, C.K.: Tidal cycle and the environmental features of Cochin Backwater (a tropical estuary). In: Proceedings of the Indian Academy of Sciences-Section B. pp. 336–348. Springer (1969)
20. Balachandran, K.K., Reddy, G.S., Revichandran, C., Srinivas, K., Vijayan, P.R., Thottam, T. J.: Modelling of tidal hydrodynamics for a tropical ecosystem with implications for pollutant dispersion (Cohin Estuary, Southwest India). Ocean Dyn. **58**, 259–273 (2008)
21. Strikwerda, M.: Cochin Estuary Morphological Modelling And Coastal Zone Management. TU Delft, Faculty of Civil Engineering and Geosciences, Hydraulic Engineering (2004)
22. Eldho, T.I., Chandramohan, P.: V: Hydrodynamics and salinity transport modelling of Cochin Estuary. ISH J. Hydraul. Eng. **11**, 163–177 (2005)
23. Paul, P., Cvetkovic, V.: Modeling hydrodynamics of real world estuarine systems. In: Proceedings of COMSOL Conference (2007)
24. Planning Commission, G. of I.: Report on Visit to Vembanad Kol, Kerala, a Wetland included under National Wetland Conversation and Management Programme of the Ministry of Environment and Forests (2008)
25. Kokkal, K., Harinarayanan, P., Sahu, K.K.: Wetlands of Kerala. In: Proceeding of Taal 2007 The 12th World Lake Conference (2008)
26. Swaminathan, M.S.: Measures to Mitigate Agrarian Distress in Alappuzha and Kuttanad Wetland Systems. A Study Rep. by Swaminathan Research Foundation, Union Ministry Agriculture (2007)
27. Sankaranarayanan, V.N., Qasim, S.Z.: Nutrients of the Cochin Backwater in relation to environmental characteristics. Mar. Biol. **2**, 236–247 (1969)
28. Lakshmanan, P.T., Shynamma, C.S., Balchand, A.N., Kurup, P.G., Nambisan, P.N.K.: Distribution and seasonal variation of temperature and salinity in Cochin backwaters. Indian J. Mar. Sci. **16**, 99–102 (1982)
29. George, M.J.: The influence of backwaters and estuaries on marine prawn resources. In: ICAR and CMFRI (eds.) Proceedings of the Symposium on Living Resources of the seas around India, pp. 563–569 (1973)
30. Azis, A., Nair, N.: Certain aspects of pollution in the aquatic ecosystems of the South west Coast of India. In: Proceedings of Seminar on Status of Environmental Studies in India, pp. 345–356 (1981)

31. Batcha, A.S.M., Damodaran, R.: Impact of Thannirmukham bund and Idukki Hydro electric project on the changes of salinity characteristics of the Vembanad Lake (South India). Arch. Hydrobiol. Beih. **28**, 193–200 (1987)
32. IAEA: Generic models for use in assessing the impact of discharges of radioactive substances to the environment. IAEA (2001)
33. Quay, P.D., Broecker, W.S., Hesslein, R.H., Fee, E.J., Schindler, D.W.: Whole lake tritium spikes to measure horizontal and vertical mixing rates. In: Isotopes in Lake Studies (1979)
34. Monte, L., Perianez, R., Boyer, P., Smith, J.T., Brittain, J.E.: The role of physical processes controlling the behaviour of radionuclide contaminants in the aquatic environment: a review of state-of-the-art modelling approaches. J. Environ. Radioact. **100**, 779–784 (2009)
35. Arnold, J.G., Srinivasan, R., Muttiah, R.S., Williams, J.R.: Large area hydrologic modeling and assessment part I: Model development. J. Am. Water Resour. Assoc. **34**, 73–89 (1998)
36. Neitsch, S.L., Arnold, J.G., Kiniry, J.R. e al, Srinivasan, R., Williams, J.R.: Soil and water assessment tool user's manual version 2000. GSWRL Rep. 202 (2002)
37. Haldar, R.: Hydrologic modelling of the eastern contributing basins of Vembanad lake using SWAT. In: 2012 SWAT Conference Proceedings. Texas Water Resources Institute (2013)
38. Priju, C.P., Narayana, A.C.: Heavy and trace metals in Vembanad Lake sediments. Int. J. Environ. Res. **1**, 280–289 (2007)
39. Selvam, A.P., Priya, S.L., Banerjee, K., Hariharan, G., Purvaja, R., Ramesh, R.: Heavy metal assessment using geochemical and statistical tools in the surface sediments of Vembanad Lake, Southwest Coast of India. Environ. Monit. Assess. **184**, 5899–5915 (2012)
40. DineshKumar, P.K., Sankaranarayanan, V.N., Devi, K.S.: Cochin backwaters: An introduction to the system, prior studies, historical trends and future implication. Indian J. Environ. Prot. **14**, 98–102 (1994)
41. DineshKumar, P.K.: Cochin backwaters: a sad story of manipulation. Ambio **26**, 249–260 (1997)
42. Remani, K.N., Jayakumar, P., Jalaja, T.K.: Environmental problems and management aspects of Vembanad Kol Wetlands in South West Coast of India. Nature, Environ. Pollut. Technol. **9**, 247–254 (2010)
43. (CPCB), C.P.C.B.: Pollution Potential of Industries in Coastal Areas of India. Coastal Pollution Control Series: Central Pollution Control Board Report (1996)
44. Harikumar, P.S., Nasir, U.P., Rahman, M.P.M.: Distribution of heavy metals in the core sediments of a tropical wetland system. Int. J. Environ. Sci. Technol. **6**, 225–232 (2009)
45. Unni, P.N., Nair, S.R.: Environmental issues in Vembanad estuary due to salinity and flood control structures. In: The 9th 1995 Conference on Coastal Zone, Tampa, FL, USA, 07/16-21/95, pp. 549–550 (1995)

Impact of Urbanization on Surface Runoff Characteristics at Catchment Scale

Manish Kumar Sinha, Triambak Baghel, Klaus Baier, Mukesh Kumar Verma, Ramakar Jha and Rafig Azzam

Abstract Increase in population and rapid urbanization are two main challenges to urban water management, especially for cities in developing countries like India. The assessment of changes in catchment surface runoff due to urbanization is critical for water resource planning and management. Uncertainty in rainfall and changing landuse pattern results from urbanization is difficult to correlate with present changing surface runoff conditions. This study has shown an approach to find a relation between these three by using a statistical term dynamic degree with application of the well-established tool, Curve Number (SCS-CN) method to observe surface runoff over the study area in conjunction with Geographic information system and remote sensing. This study assesses changes in runoff characteristics of Raipur catchment which has been delineated from a smallest sub-basin of Mahanadi River in Chhattisgarh, India. Pixel-based Runoff depth of each sub-catchment was estimated by incorporating digital elevation model, rainfall data, and Land Use/Land Cover (LULC) information for the years 1971–2015. Spatio-temporal variation in each sub-catchment has been computed and significant change in runoff has been observed. The motivation of this study aims to understand and to study the changing nature of urban growth pattern/sprawl, to quantify surface imperviousness and changes in surface runoff characteristics.

Keywords Landuse land cover · HEC-HMS · SCS-CN · Surface runoff Urbanization

---

M. K. Sinha (✉) · K. Baier · R. Azzam
Department of Engineering Geology and Hydrogeology, RWTH Aachen University, Aachen 52064, Germany
e-mail: sinha@lih.rwth-aachen.de

T. Baghel · M. K. Verma
Department of Civil Engineering, National Institute of Technology Raipur, Raipur 492010, India

R. Jha
Department of Civil Engineering, National Institute of Technology Patna, Patna 800005, India

M. Rathinasamy et al. (eds.), *Water Resources and Environmental Engineering I*,
https://doi.org/10.1007/978-981-13-2044-6_3

1 Introduction

To adequately manage impacts of ongoing or future land use changes, knowledge about existing landuse and trends of change is essential to tackle the problems associated with haphazard and uncontrolled growth. It is evident that land and water, an important life supporting systems are under intense pressure due to human induced accelerating factors [1]. In 2008 the world reached the historical mark of more than half of world's people living in cities than in rural areas [2]. The increasing urban population is expected to grow further, that two-third of the total world's population will be urban habitant until the year 2050 [3]. India holding the second largest population in the world, in terms of an aerial increase, from the year 2005 cities all over the world will expand on average 2.5 times in size till the year 2030 [4]. This increasing rate of urbanization is mostly expected in developing countries [5]. This increase in urban population will cause a huge infrastructure demand that will lead to change in a significant amount of natural lands to urban infrastructure. These urban infrastructures are a kind of scourge to mankind with in this century as it is directly related to urban imperviousness. It is no more covered that increasing the built-up landuse is the cause of less groundwater recharge and high surface runoff in an urban area.

The study of spatio-temporal characteristics of an urban sprawl and land use pattern is useful for the sustainable land-management and urban-land planning [6, 7]. After some significant literature review, USGS earth observational satellite data (Landsat) are found to be best suitable for this study. Hence, landsat imageries were used for the LULC classification. This paper covers LULC analysis of an urbanized catchment using ERDAS Imagine software and rainfall-runoff simulation using HEC-HMS modeling platform for six consecutive time periods has been taken for Raipur catchment. It has been seen that due to high urban growth and building of informal/unrolled structures the surface imperviousness is increasing which is directly affect the runoff condition in the catchment [5]. Hence a huge changing pattern would be seen in groundwater recharge and flooding in urban area. Maximum likelihood classifier has been used to perform supervised classification with satellite imageries. Later the LULC maps were used to generate curve number (CN) values and percentage imperviousness values for the catchment. This is again used as an input in HEC-HMS model to create continuous rainfall runoff simulation for different years.

2 Materials and Methods

Present study is focused on a Raipur city in India. As far as the study concern with catchment scale, a small catchment has been delineated from available DEM (Digital Elevation Model) based upon the one inlet and one outlet point. The above catchment can be seen in Fig. 1. The delineated Raipur catchment area is 1118 km^2

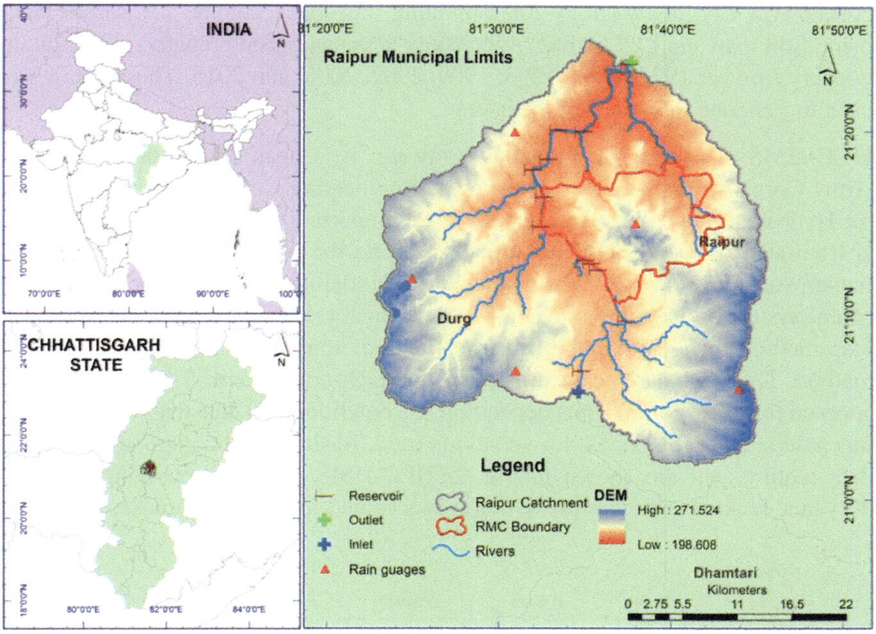

Fig. 1 Location map of the Raipur catchment (designated delta shape is Raipur catchment and red boundary is Raipur municipal boundary)

which is located between 21°00′ N and 21°30′ N latitude and 81°20′ E longitude. The whole of Raipur catchment is part of Kharun sub-basin, which is thickly populated and closely cultivated [8]. The climate of the study area is sub-tropical with three distinct seasons. Temperatures remain moderate for most of the year. May is the hottest month and December is the coldest. The city experiences the average annual rainfall of 1200 mm. The Kharun River flows to the north and located at east of the Raipur city, and passes through the center from the Raipur catchment, which is also a source of domestic water supply to the city [9–11]. All the analysis is performed in the catchment scale using HEC-HMS but the present paper covers the anomalies only inside the Raipur municipal corporation area (shown in Red color) which has approximately 300 km^2 area. The city has a population of 1.125 million and experienced a growth rate of 51.06% during decade 2001–2011 [12].

The relationship between rainfall-runoff (RR) is very complex and influenced by various storm event and drainage basin characteristics. RR can be estimated using several approaches. CN (Curve Number) based SCS (Soil Conservation Service-Curve) method is well established for dynamic RR modeling for assuming the catchment as semi-distributed model [13]. SCS-CN method has been developed by National Resources Conservation Service (NRSC), United States Department of Agriculture (USDA) in 1969 that is simple predictable and stable conceptual method for computation of direct runoff based on rainfall depth [14]. Present study covers SCS-CN methodology for estimating the runoff volume for Raipur catchment. Soil map and LULC map of the study area have been prepared using satellite imagery for year 1972, 1989, 1993, 1999, 2002, 2008, and 2015. These data's were used to generate CN map for each year.

HEC-HMS is hydrologic modeling software. It is open source product of U.S. Army Corps of Engineering's Hydrologic Engineering Center (HEC). HMS stands for Hydrologic modeling system. One of the important model components of HMS is Rainfall Runoff Model. This model converts the precipitation data in the particular watershed to runoff. For mathematical modeling purpose, complex sub-basin characters are divided into five components which are modeled separately. They are loss method, canopy method, surface method, transform method and baseflow method. Different methods are available in each component. The methods to be adopted for those components depend on site condition and data availability [15]. In this assessment HEC-HMS 4.5 version is used. Basic Model components and their data requirement are shown in Fig. 2. HEC-HMS has been evolving over past 30 years; recent researches prove that it is accepted for official purposes also. Since

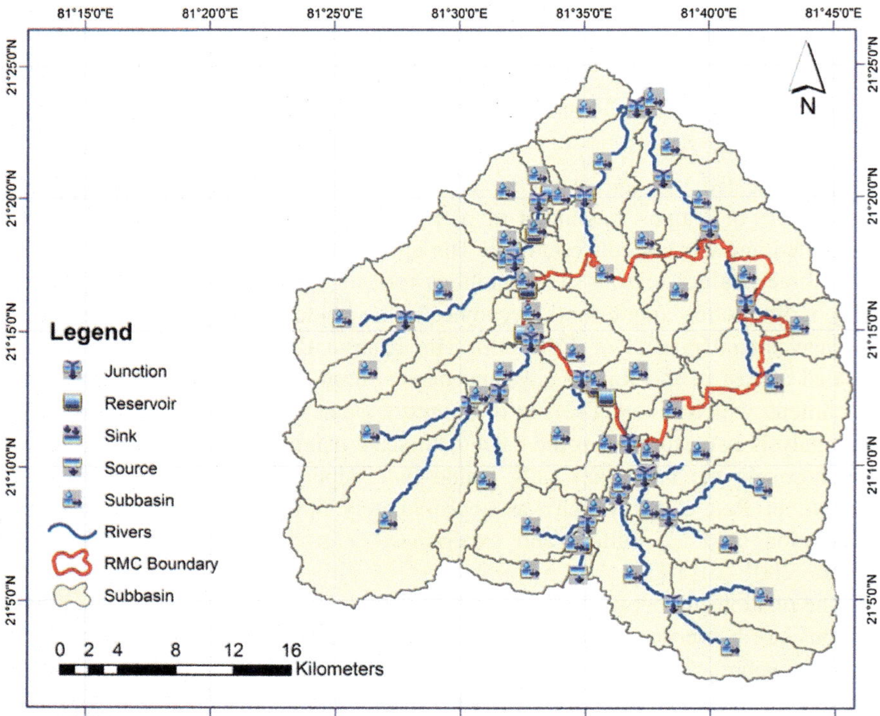

Fig. 2 HEC-HMS model setup with HEC schema of Raipur catchment used for simulation for 47 subdivided sub-basins including 12 anicuts operation into main river

Table 1 Process, input data used to generate HEC-HMS model for Raipur catchment

Process	Method used	Input data	Equation used
Modeling runoff volume	Grided SCS CN method	Initial abstraction, curve number, percent impervious	$Q = \frac{(P-\lambda.S)^2}{(P-\lambda.S+S)}$
Modeling direct runoff	SCS Unit hydrograph model	Basin lag	$\text{Lag\_time} = \frac{I^{0.8} \times (S+1)^{0.7}}{1900 \times Y^{0.5}}$
Modeling base flow	Recession	Observed discharge	$Q_t = Q_0.k^t$
Modeling channel flow	Muskingum Kung	Mannings n, Shape (8 point method)	$S_t = K[XI_t + (1 - X)O_t]$

Where,
P = daily rainfall in mm
λ = In the SCS method in Indian condition the NIH, 1998 gave the value of λ coefficient equal to 0.2, although for black soil region S = 25,400/CN − 254
I (m) = Hydraulic length of watershed; Y = Avg. land slope
Q_0 = initial base flow (at time zero) in m^3/s; k = an exponential decay; K = travel time of flood wave and X = dimensionless weight (0<X<0.5)

the software has many options in each method, it is flexible for different site conditions. The methods/process applied in this study is denoted in Table 1.

Figure 2 shows a typical model setup in HEC-HMS of Raipur catchment. All the model processes are assigned in HEC-Geo-HMS and input using HEC-DSS tool.

3 Results and Discussion

Result of supervised classification of landuse land cover of Raipur city area has been shown in Table 2, based on pixel size it shows the total area cover by each landuse classes for each of the year. The results clearly show that urban areas have expanding and water bodies are decreasing since 1989 with continuous population growth. Rapid socio-economic development, huge opportunity of business and market activity inside the central part of the city, such rise in urban area is expected. In other hand agriculture land and vegetation land has random anomaly in different years. Due to coarser resolution imagery it can be consider that the agriculture land and vegetation land has not been recognized at the time of visual interpretation. Also the Raipur city is the most polluted city in India in terms of air pollution [16]. This is also a reason for lack of visual interpretation and histogram error in north side of the city where industrial area is populated.

At the stage of development the landuse category of agriculture and open land area were converted into urban areas. Also in the same way, the landuse types which has lost the spatial extent is open land and agriculture landuse type that reduced an area of 27.62 and 25.82 km^2 respectively to other landuse types. Since 1972–2015 in Raipur city has witnessed significant changes in Water bodies, Open land and Agriculture area due to continuous urban encroachment (Fig. 3).

Table 2 Landuse Landover classification using supervised classification in ERDAS for different satellite products and calculated dynamic degree in percentage

Sat. Sensor	MSS	TM			ETM+			Area change in 1972–2015	Dynamic degree in % 1972–2015
Year/LULC	1972	1989	1993	1999	2002	2008	2015		
Agricultural	199.4	92.9	164.3	214.1	184.2	137.1	173.5	−25.82	0.30
Open land	44.3	125.7	67.2	42.2	34.9	28.3	16.7	−27.62	1.45
Urban	3.9	6.3	7.5	9.12	12.9	37.5	53.4	49.49	−29.51
Vegetation	48.9	71.8	58.1	32.0	65.2	95.0	54.3	5.44	−0.26
Water Bodies	3.8	3.8	3.53	3.2	3.4	2.88	2.5	−1.36	0.82
Total	300.4	300.7	300.7	300.7	300.7	300.7	300.5		

Note All values are in km^2

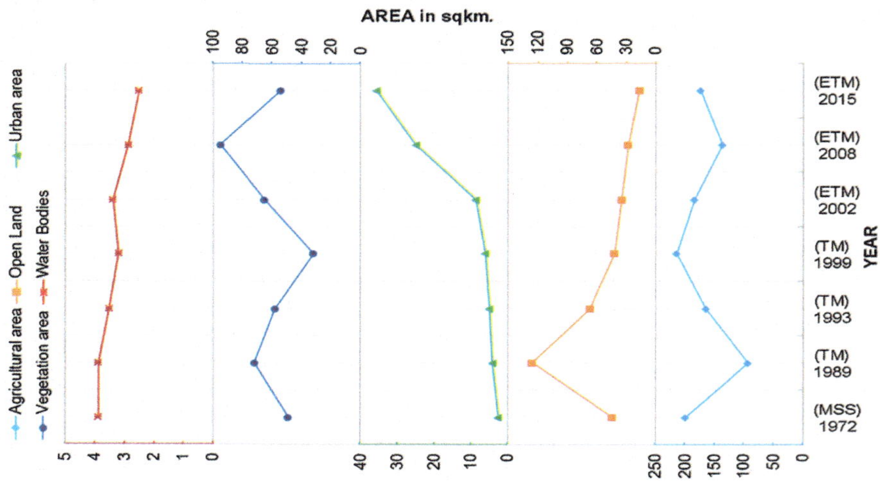

Fig. 3 Temporal variation in landuse landcover classification of Raipur city

To compare this anomaly mathematically one equation has been used in similar study by Wakode, Baier [7]. The changes in landuse in a temporal extent can also be calculated using the dynamic degree equations. Table 2 shows the dynamic degree of landuse types that is a quantitative variation of a landuse category with in a time scale and can be calculated using [7]:

$$K(\%) = (Z_a - Z_b)/Z_a \cdot (1/T) \cdot 100,$$

where
K = the dynamic degree change in a particular landuse type; Z_a and Z_b = the area of a type of landuse before the change and after the change, respectively; T is the time interval between two successive year of landuse change.

Instead presenting change in percentage area, dynamic degree has been computed which has a time factor. Negative values in dynamic degree shows higher growth in land use change and vice versa. Using the resulted landuse classification and soil classification curve number map has been generated. This is further used for input for HMS model as Basin CN and percentage imperviousness. The above shown HEC-HMS model in Fig. 2 has been run and simulate for seven different control specifications to compute runoff at each pixel length as it is mentioned the gridded SCS CN method has been implemented for present study. The purpose of selection of grid based method for calculation is because of RMC boundary which is different than sub-basin boundaries in Raipur catchment model. This approach reduces the limitation of composite CN, instead using grid-based method in this study. Daily rainfall data has been used to delineate the runoff quantity in the catchment which is summarized in Table 3 as monthly and distribution curves for runoff and bar diagram for rainfall has shown in Fig. 4.

From this continuous rainfall runoff simulation in Table 3 the annual rainfall is decreasing so that runoff is also decreasing. In Table 3 two variables that are rainfall and CN/year of landuse. Both are changing means it is difficult to interpretate the changes in runoff due to urbanization because landuse is also changing as in the same time rainfall patterns changes which is natural in nature. To draw or conclude result from the present analysis a change in dynamic degree between rainfall runoff and urban-landuse has been computed (Table 4). The annual rainfall and runoff has been successfully correlated by commuting correlation coefficient which is varies from 0.76 to 0.86. Three new parameter is calculated which includes time duration of landuse change are Rainfall, runoff and urban landuse change as % dynamic change. To correlate the change in landuse with runoff the values were plotted in line graph type against duration (Fig. 5). The results shows that the runoff is follow the exact urban landuse distribution which means dynamic degree change in runoff is same as urban landuse change. It may happed due to runoff calculation in this method is dominated by CN values which is directly related with landuse. So it can be concluded that as urban land is increasing the runoff trend is also increasing.

4 Conclusion

This study exhibit the use of RS and GIS techniques, to characterize the process of urbanization efficiently. The freely available historical MSS, TM and ETM + sensor data along with census records have helped in identification of LULC change in Raipur city. It indicated that total urban area has grown since 1972–2015 by 49.49 km^2. Open land and agriculture landuse types are mostly affected by the urban growth. Reduction of water bodies since 1972 were significant and most of the remaining water bodies shows further reduction of its spatial extent. The total number of registered lakes in Raipur city was 185 in numbers now only 85 has been left. Such losses of surface water resources are also one of the reasons in changes of surface runoff pattern in Raipur city. Result of HEC-HMS model reveals

Table 3 Monthly total rainfall and generated runoff at the outlet of Raipur catchment (Runoff represented in mm as equality distributed through the basin, to calculate exact runoff volume the runoff depth must be multiplied by corresponding sub-catchment area)

	1972		1989		1993		1999		2002		2008		2015	
Months	Rainfall (mm)	Runoff (mm)	Rainfall (mm)	Runoff (mm)	Rainfall (mm)	Runoff (mm)	Rainfall (mm)	Runoff (mm)	Rainfall (mm)	Runoff (mm)	Rainfall (mm)	Runoff (mm)	Rainfall (mm)	Runoff (mm)
Jan	0	0	0	0	0	0	0	0	13.8	41.2	7.6	19.2	9.4	27.5
Feb	5	10.4	0	0	10.2	18.7	11	30.8	5.2	25.1	11.6	50.9	2.2	9.3
Mar	0	0	12.6	34.4	18.5	99.2	0	0	5.7	30.9	24.8	150.1	19.3	106.9
Apr	8.6	28.6	0	0.2	14.2	88.4	0	0	3	17.4	5.2	34.5	51.4	350.5
May	21.2	119.5	12	59.8	32.8	214.5	70	447.4	14.1	86.5	0.6	4	13.4	95.8
Jun	213.2	1503	457.1	3288	135.4	985.9	98.6	696.8	196	1404	204	1448	271.6	1989
Jul	199.6	1546	391.1	2909	499.8	3688	135.9	1016	71.8	492.2	125	940.1	173.2	1282
Aug	374.3	2790	227.7	1649	405.2	2991	421	3031	345	2571	229	1628	267.4	1971
Sep	192.4	1484	196.2	1488	166.5	1238	263.7	2030	90.2	676.5	220	1693	219.6	1631
Oct	22	147.1	0	0	45.3	335.1	21	155.8	23.2	171.5	12.6	93.2	0	0
Nov	51	403.4	0	0	0	0	0	0	0	0	0	0	0	0
Dec	2	16.9	13.6	99.5	0	0	0	0	0	0	0	0	13.8	101.1
Total	1089	8049	1310	9528	1328	9659	1021	7407	767	5516	840	6061	1041	7564

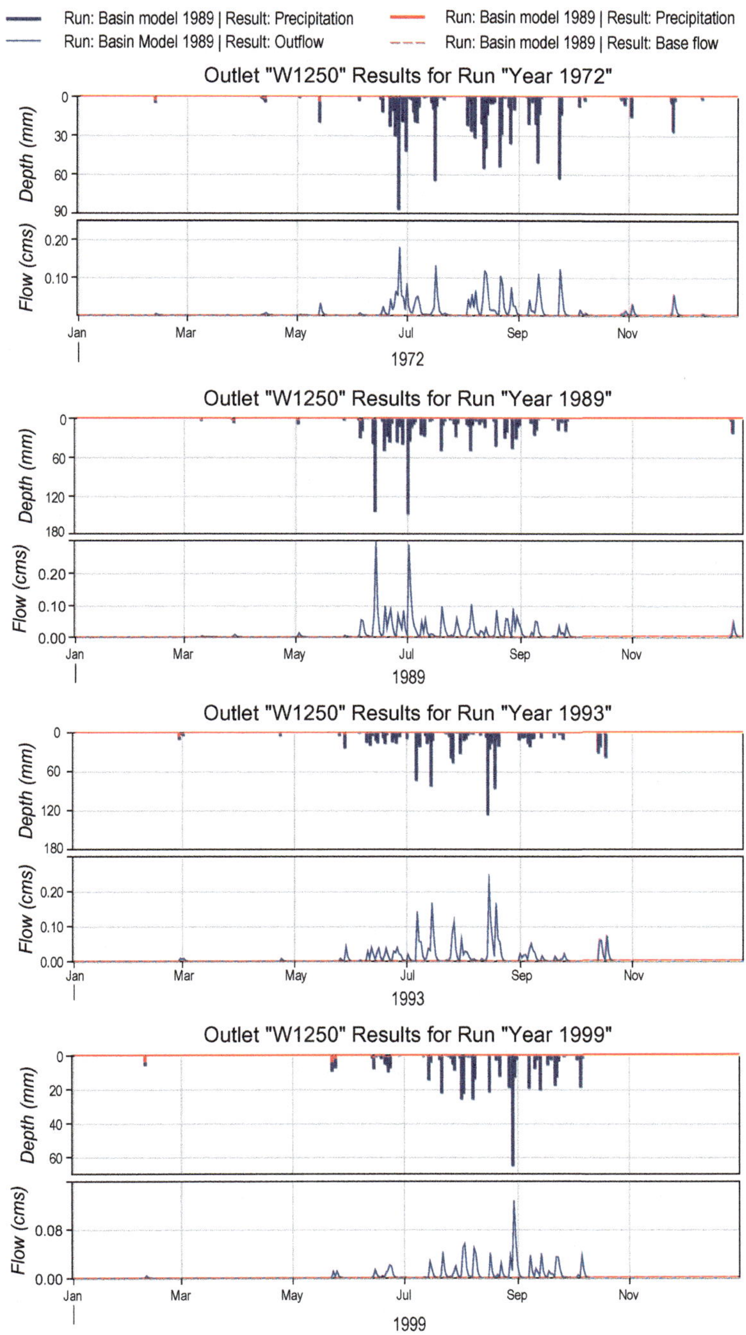

Fig. 4 HEC-HMS Simulated Rainfall (top), Runoff (Bottom) and losses at Sub-basin 'W1250'—catchment outlet form 1972 to 2015 in chronological order

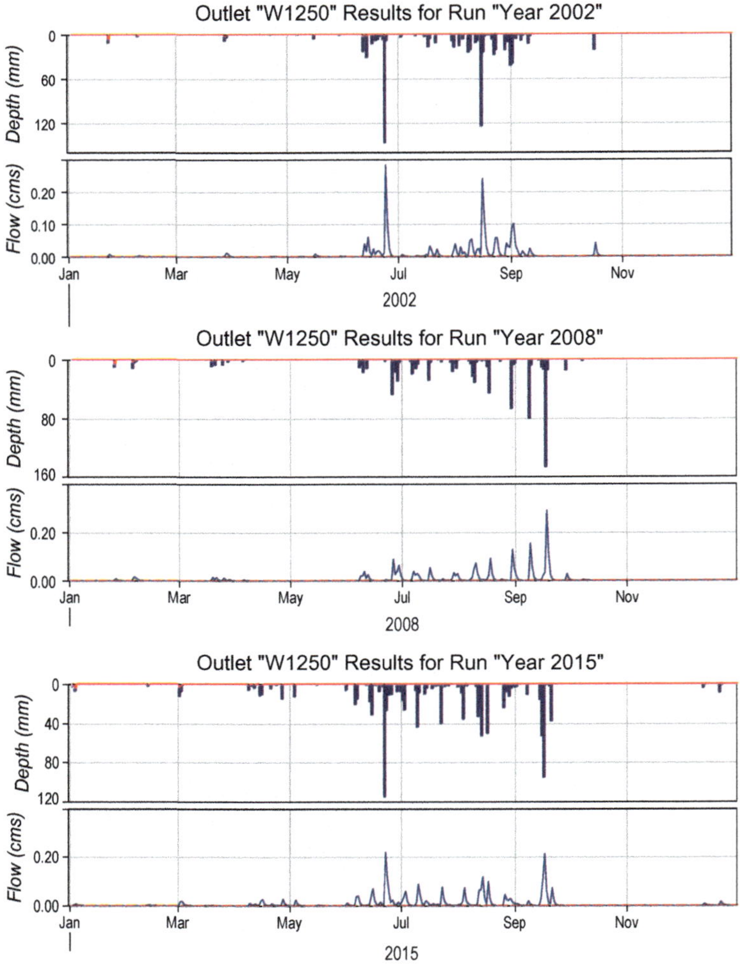

Fig. 4 (continued)

that the landuse change is a most important factor for changing pattern in runoff in Raipur city. The simulation result of HEC-HMS model can also be used for forecasting the future changes in runoff pattern in the city if proper calibration has been applied. This study produces very valuable results for planner/policy makers as it reveals a very comprehensive indication of surface runoff distribution in Raipur city. The results can be conveniently utilized by the water resource engineering decision-makers for scientific management planning to mitigate the quantity of water resources. Moreover, it provides a sound base for planning of land use practices.

Table 4 Rainfall runoff correlation and dynamic degree change with urban landuse

Year	Annual rainfall in mm	Annual runoff in mm	Urban landuse in km^2	Dynamic degree in %			Correlation between rainfall and runoff	
				Rainfall	Runoff	Urban landuse		
1972	1089.3	8048.7	3.9	–	–	–	$y = 7.5069x - 10.717$	$R^2 = 0.82$
1989	1310.3	9528.2	6.3	−3.62	−1.19	−1.08	$y = 7.3315x - 6.5197$	$R^2 = 0.86$
1993	1327.9	9659.1	7.52	−4.84	−0.34	−0.34	$y = 7.4213x - 16.299$	$R^2 = 0.79$
1999	1021.2	7407.3	9.12	−3.55	3.85	3.89	$y = 7.3506x - 8.2597$	$R^2 = 0.81$
2002	767.2	5516.3	12.98	−14.11	8.29	8.51	$y = 7.4469x - 16.413$	$R^2 = 0.76$
2008	840.2	6061	37.5	−31.48	−1.59	−1.65	$y = 7.381x - 11.711$	$R^2 = 0.79$
2015	1041.3	7563.5	53.39	−6.05	−3.42	−3.54	$y = 7.423x - 13.836$	$R^2 = 0.86$

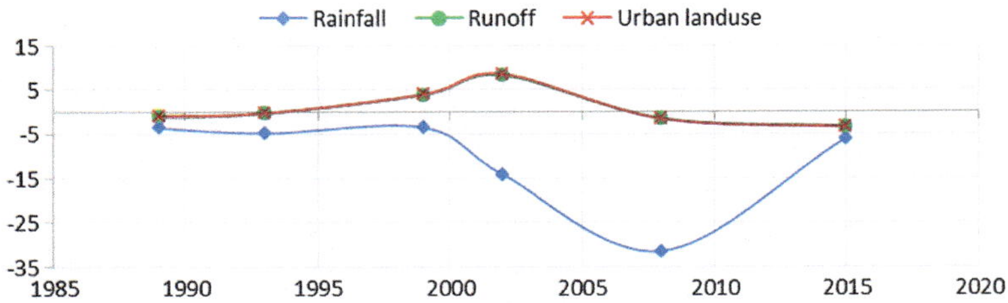

Fig. 5 % Dynamic degree change between Rainfall, runoff and urban landuse

Acknowledgements The authors would like to thank the Deutscher Akademischer Austauschdienst-German Academic Exchange Service (DAAD), as this study was carried out within the program "A New Passage to India".

References

1. UNICEF: Water in India: situation and prospects, in UNICEF, India Country Office (2013)
2. Un, D.: World Urbanization Prospects: The 2014 Revision. United Nations Department of Economics and Social Affairs, Population Division, New York, NY, USA (2015)
3. Un-Habitat: Planning Sustainable Cities: Global Report on Human Settlements 2009. Routledge (2016)
4. Angel, S., et al.: The Dynamics of Global Urban Expansion, Transport and Urban Development Department. World Bank, Washington, DC (2005)
5. Montgomery, M.R.: The urban transformation of the developing world. Science **319**(5864), 761–764 (2008)
6. Strohschön, R., et al.: Land use and water quality in Guangzhou, China: a survey of ecological and social vulnerability in four urban units of the rapidly developing megacity. Int. J. Environ. Res. **7**(2), 343–358 (2013)
7. Wakode, H.B., et al.: Analysis of urban growth using Landsat TM/ETM data and GIS—a case study of Hyderabad, India. Arab. J. Geosci. **7**(1), 109–121 (2013)
8. Sinha, M.K., et al.: Assessment of groundwater vulnerability using modified DRASTIC model in Kharun Basin, Chhattisgarh, India. Arab. J. Geosci. **9**(2), 1–22 (2016)
9. GIZ, ASEM, and RMC: Annexures City Sanitation Plan for Raipur (2011)
10. RMC: Detailed Project Report, Sewerage, Storm, Surface water drains and lake protection works for Raipur city. Meinhardt (Singapore) Pte. Ltd (2010)
11. RMC: Raipur City: Storm Water Drainage Report (2011)
12. Census-of-India: Census of India 2011, District census handbook: Raipur, pp. 1–452 (2011)
13. Nigam, G.K., et al., Field assessment of surface runoff, sediment yield and soil erosion in opencast mines in Chirimiri area, Chhattisgarh, India. Phys. Chem. Earth, Parts A/B/C (2017)
14. Jena, S.K., et al.: RS and geographical information system based evaluation of distributed and composite curve number techniques. J. Hydrol. Eng. **17**(11), 1278–1286 (2012)
15. Feldman, A.D.: Hydrologic modeling system HEC-HMS: Technical reference manual. US Army Corps of Engineers, Hydrologic Engineering Center (2000)
16. Verma, M.K., et al.: Computation of air quality index for major cities of Chhattisgarh state. Environ. Claims J. **28**(3), 195–205 (2016)

Special Considerations for Design of Storm Water Drainage System—A Case Study

Kuppili Rajeswara Rao and Ponnada Markandeya Raju

Abstract Urban Floods are becoming more frequent owing to the lack of proper storm water drains. In this paper we discuss some of the important design features that need to be considered for design of storm water drainage system using a real-world case study. The flood events during September–December 2010 in the southern side of the Visakhapatnam city (INDIA) were considered. The recommendations are evolved through ground reconnaissance survey, hydro-meteorological and hydrological studies.

Keywords Storm water drain · Inundation · Encroachments · Water hyacinth, etc.

1 Introduction

Unevenly distributed high-intensity rainfalls coupled with draining canals of inadequate capacity are the common causes of urban flooding. Researchers predict that rapid urbanization with poor town planning aggravated by climate changes are the most important cause the inundation of the urban areas. The main causes of Urban Flooding are Heavy Rainfall/Flash floods/coastal floods/river floods, Lack of Lakes/ponds, Siltation of drains, Siltation of water bodies and other Natural Causes. Human interventions like Deforestation, Constructions on existing natural storm water drains, Unplanned urbanization, increased paved area and decreased agricultural land (acting as a percolation zone) and Clogging of water carriers due to dust, garbage chocked gully gratings, bell mouths of roads and inlets of street drains, passing of cables, pipes across the drains, Garbage dumping in or on drains, etc., also result in Urban flooding. The flooding affects every section of people/

K. Rajeswara Rao · P. Markandeya Raju (✉)
Maharaj Vijayaram Gajapathi Raj College of Engineering (A),
Vizianagaram 535005, Andhra Pradesh, India
e-mail: markandeyaraju@gmail.com

K. Rajeswara Rao
e-mail: drkuppili@gmail.com

M. Rathinasamy et al. (eds.), *Water Resources and Environmental Engineering I*,
https://doi.org/10.1007/978-981-13-2044-6_4

systems in a city in the form of damage to Public buildings, Public utility works, housing and house-hold assets, loss of earning in industry and trade, loss of earning to petty shopkeepers and workers and employment to daily earners, loss of revenue due to Road, Railway Transportation Interruption resulting in high prices for essential commodities and cost of disaster mitigation.

Visakhapatnam is one of the fastest growing metropolitan cities in Asia. The industrialization and the accompanying urbanization are responsible for the rapid growth of the city. Sheelanagar is located at Latitude 17°43′ 16.2 and Longitude 83°13′ 28.6 in the city limits of Visakhapatnam. One irrigation tank "Krishnamraju tank" in the village limits of Nathayyapalem/Thungalam and the surplus water of this tank passes through the Sheelanagar area and finds its way through the existing culvert on NH-5 before joining the Bay of Bengal through GAIL compound. The surplus course which was intended to take care of the surplus discharge of "Krishnamraju tank" was subjected to human intervention and hence was not serving its intended purpose. This was identified as one of the root causes for inundation of the area during the monsoon season of 2010 and the effect was felt to greater extent at the end of the monsoon season. Apart from the rainy season, the Visakhapatnam District received abnormal rainfall even in the months of November and December, 2010 with particularly heavy spells of rain in and around Visakhapatnam town (as recorded by Visakhapatnam Airport and Waltair Observatory (India Meteorological Department's (IMD) Cyclone Warning Centre).

With the unprecedented rainfall in the month of December, the low-lying area of Sheelanagar got submerged and the people residing in the area were subjected to a lot of inconvenience in view of stagnation of water for a period of more than 3 days. The internal drains which were silted up with the soils transported by the runoff actually dwindled the discharge capacity of the drains. This had compounded the woes of the people residing in the area. The stagnation of water for a continuous period of 3 days had also resulted in considerable damage to cement concrete internal roads as well as to the existing cement drains. Apart from this, some of the layouts are not provided with storm water drainage network. This had caused the flood waters to travel as sheet of water on the existing ground thereby compounding the miseries of the people in these layouts. In general, the city of Visakhapatnam is developed on the hill slopes. The human interference has affected the natural formation of the landscape. The contribution of silt and sand from hill slope has been intercepted by reservoir, railway lines, roads, etc. Low-lying areas in the region had not been fully provided with proper drainage system. Natural level between National Highway-5 (NH-5) and Visakhapatnam Harbour varies between +4.0 and −2.0 m with respect to Chart Datum (CD). Incidentally Sheelanagar had been developed in this low-lying area and it is obvious that in the event of heavy rainfall, the area gets inundated. Further, Sheelanagar is surrounded by two man-made interventions: Railway track in embankment on one side and NH-5 in embankment on the other side. These man-made interventions had compounded the problem.

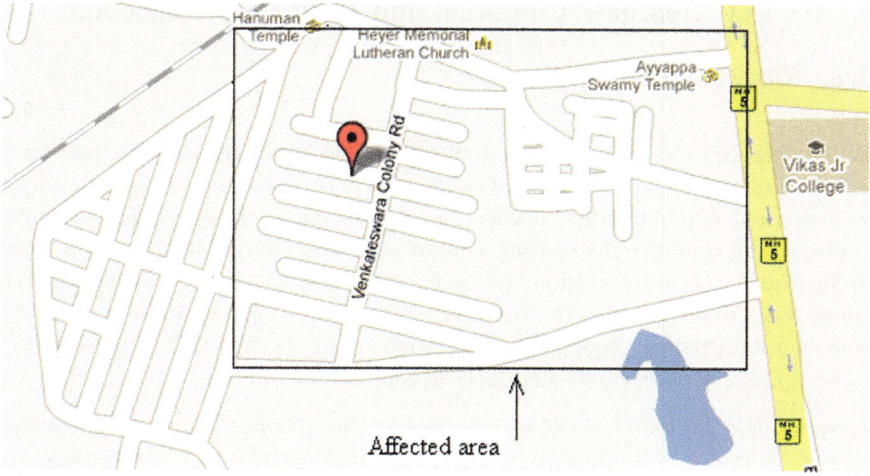

Fig. 1 Parts of Sheela Nagar which get inundate during heavy rain fall

Figure 1 shows Krishnamraju tank along with existing surplus course and the area affected during the recent flood. (including the existing storm water drainage courses in some of the layouts of Sheelanagar and the other private layouts in the vicinity that were subjected to inundation).

The steps involved are Reconnaissance of the site, problem appreciation and preliminary studies based on primary Hydro-meteorological, Hydrological, and broad topographic details for carrying out detailed investigation and submission of report.

i. Topographic and Hydrographic Surveys.
ii. Flood hydrology of the surplus of "Krishnamraju tank" based on design storm.
iii. Study of Dry Weather Flow (DWF) of existing drains with reference to designed discharge.
iv. Study of existing surplus course carrying capacity and effect of dry weather flow on the hydrology of surplus course once DWF of drains are diverted to surplus course.
v. Measures for the safe passage of flood discharge to the other side of NH-5.
vi. The design flood and critical section will be negotiated to find optimum solution.
vii. Suggestion of measures to channelize the surplus discharge of "Krishnamraju tank" along with the DWF of drains to give relief to frequently inundated area from critical flood.

2 Project Area, Site Condition and Field Data Collection

2.1 Site Conditions

With reference to Sheelanagar lay out, there is a tank viz "Krishnamraju tank" in the village limits of Thungalam and the existing surplus course for carrying the surplus waters could not fulfill this function in view of encroachments into the surplus course of the tank, thereby causing a lot of inundation during the monsoon season in Sheelanagar area. Apart from this, some of the layouts are not provided with any storm water drainage network. This caused the flood waters to travel as a sheet of water on the existing ground thereby compounding the miseries of the people in these layouts. The necessary field data is collected as follows.

Task 1: Field Reconnaissance and Secondary Data Collection A reconnaissance survey of project was done to get a better understanding of the project area. Secondary information required to design the proposed surplus channel from Krishnamraju tank and storm water drainage system was collected from urban local bodies, GVMC and from other relevant governing agencies. The objectives of the collection of data is to be familiarized with urban characteristics of the city and its environment, and obtain first-hand information on the land use, topography, soil characteristics, existing drainage system, etc. The documents and data sources that were collected and reviewed include the following as listed in the Table 1. The

Table 1 Documents and Data Source

Component	Document
Land use and Development Trends	Master Plan Report (present and proposed), Land use Map, Satellite Imageries/Ariel Photographs, (subject to availability with client) Development Control Rules
Demographic and Economic Characteristics	Census (Directorate of Economics and Statistics), Industrial Profile
Top sheets and Contour maps	Corporation/UDA, Survey of India
Infrastructure like, Water Supply, Sewerage, Roads, Sanitation, Street lighting, Storm water Drainage and Solid Waste	Documents, Maps, Data and Reports from Corporation, PHED, Irrigation, PWD etc.
Rainfall & Meteorological trends, Isohyte Maps	Meteorological Department
Traffic & Transport Infrastructure	Corporation/UDA, Roads & Buildings Department, Railways
Soil Characteristics and Hydrology	Corporation/UDA, Agricultural and Ground Water Department
Informal settlements & encroachments	Documents from Corporation, Studies carried out as part of slum up-gradation program
Opens spaces and Natural Water Bodies	Information from Corporation/UDA, Characteristics of water bodies from PWD/ Irrigation departments

main objective of this task is to establish development trends in the study area, and prepare an inventory of existing information in the above aspects. The review helped to identify the physical constraints to drainage, both natural and man-made.

Task 2: Review of the Existing Storm Water Drainage System The existing drainage system was studied in detail to identify the deficiencies and to act as a base for designing the proposed system. The review includes mapping existing drainage zones, characteristics of primary and secondary drains, existing disposal practices and arrangement, identification of low-lying and flood-prone areas, identification of physical and natural constraints (including culverts, bridges, canals, transport infrastructure).

Task 3: Topographic Surveys This task would entail carrying out topographic surveys, the key input in the planning and design of the storm water drainage system. The surveys would be carried out using total station instrument and other equipments and comprises of the following:

- Identification of Benchmark established by Survey of India (SOI) in project area.
- Establishment of network of permanent bench marks.
- Study of the Survey of India topographical maps for understanding of the topography, water courses and other physical features like major roads, railway lines, location and level of GTS bench marks, etc.
- Topographic survey along the roads and levels taken approximately at 25 m c/c for preparing longitudinal sections and contour map.

In addition, a condition assessment of the existing major drain was carried out to ascertain the physical, structural, and hydraulic conditions. This would entail assessing the extent of encroachments, siltation and pollution of drains due to disposal of solid waste. Data from Topographic and Hydrographic survey, details of storm water drains and culverts in the layout. Further, the maximum flood levels among various layouts during the recent inundation were also observed.

2.2 Scope of Work for Field Data Collection

The scope of work includes the following. Establishment of bench mark: Establishment of Bench mark to be carried out as per Survey of India specifications and connecting the same to GTS Bench Mark and topographic survey.

2.3 Field Data Collection and Analysis

Area of Survey The area of survey is located surrounding Sheelanagar and the colonies in the vicinity and the catchment area of Krishnamraju tank of Tungalam (V). The map of Visakhapatnam along with the location of the inundated/affected area along with railway lines and important Roadways are shown in Fig. 1. The catchment area of Krishnamraju tank in Nathayyapalem (V) is obtained from the respective Topo sheet. Total station instrument was used for the entire surveys covered in this work.

Analysis of Field Data (Topographic and Hydrographic Survey) The levels measured are with reference to Mean Sea Level. The L.S and site plan showing the alignment of the proposed surplus channel from Surplus weir to GAIL compound wall on the other side of NH-5 is furnished in Fig. 2.
During heavy rains, Krishnamraju tank attains full tank level and the excess water is released into the surplus course. The bed level is taken as +6.05 m, being 1.80 m below the apron level on downstream side of existing surplus weir of tank which is +7.85 m with reference to MSL. The top of bank level was taken. Cross sections were taken at 25 m interval from this section as zero chainage. The surplus course was encroached at certain points and detoured at certain points. Apart from topographic survey and hydrographic survey studies, the details of DWF of the residential drains in the area are also collected, which would be required for earmarking the problematic areas for the design of flood control measures.

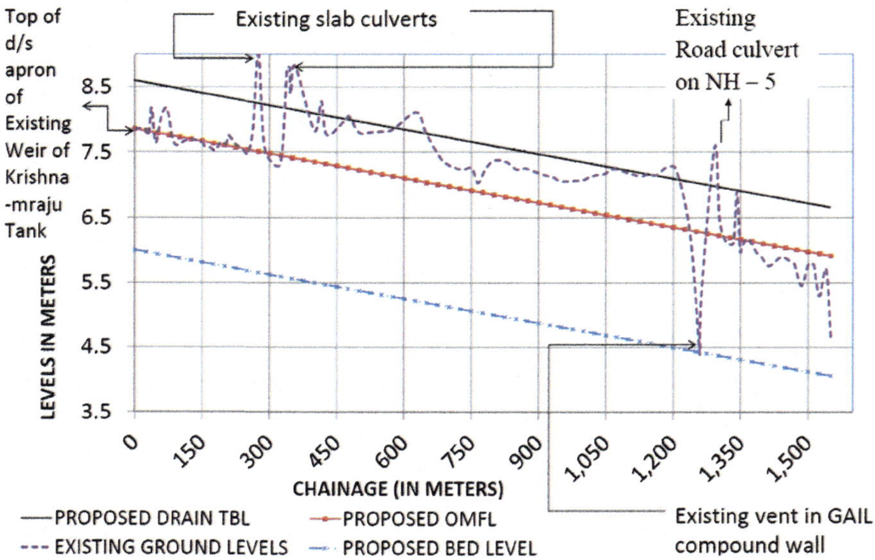

Fig. 2 Existing ground level and proposed drain levels

3 Hydro-meteorological Studies

3.1 Overview of Visakhapatnam

Topography Visakhapatnam Corporation altitude is 5 m above MSL. It is general topography is gently sloped towards Bay of Bengal. Due to this, the city and its environs can be divided into four categories viz. Hilly region, Upland tracks, Rolling plains and Plains. The area of the city is 111.61 km^2 and is divided into 72 wards as six zones with large industrial base (Visakhapatnam Steel Plant, the Bharat Heavy Plate and Vessels and the Hindustan Zinc.

(a) Climate and rainfall: Visakhapatnam has hot (up to 45°) climate during summers (typically from March to May), pleasant winters (with temperature between 18 and 32 °C during December to February) and heavy rainy monsoons during June to September. Being close to the sea, the level of humidity is high. The average annual rainfall for the city and its surroundings is of the order of 950 mm with the bulk of the rain coming from northeast monsoon. During the period between September–November, storms and depressions originating in the Bay of Bengal cross the east coast in the neighborhood causing heavy rains and gales. The details are furnished in Fig. 3.
(b) Geology and groundwater table

The soil and subsoil in the region is mostly clay and sandy/silty in nature. Being a coastal area the groundwater table level is as high as +0.5 to +3 m. The level is likely to go down by 1 m during summer. As the area is a part of marine tidal flat region, the ground water is saline. Ground water in the basin is utilized for domestic, irrigational and industrial purposes.

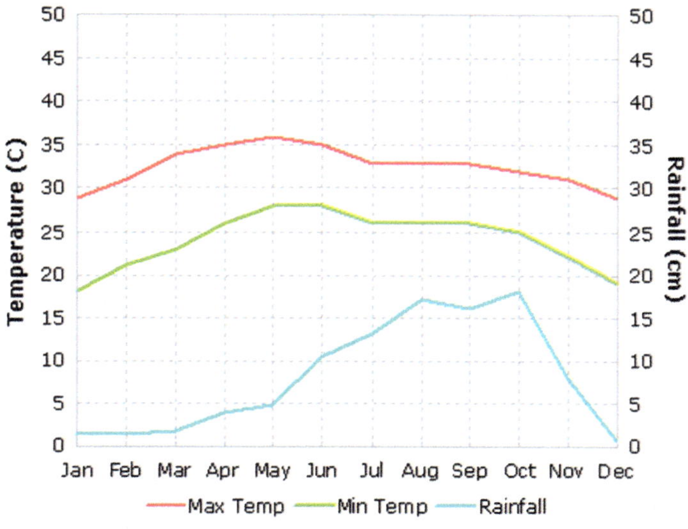

Fig. 3 Normal variation of temperature and rainfall among various months

3.2 An Appraisal of Daily Rainfall Data

It is necessary to study the rainfall data of many earlier years to find the magnitude of the storm water runoff (portion of the precipitation which drains over the ground surface and join the storm sewers/nallas) in the area of concern. A storm water drainage system is a system for proper collection and conveyance of this rainfall precipitation to the corresponding disposal point. Intensity and duration of precipitation, characteristics of the catchment area and contributing storm runoff are some factors affecting the magnitude of storm water runoff. As the intensity of rainfall and its duration are not uniform over a period, the rainfall precipitation with the duration equal to or more than the time required for overland flow needs to be analyzed for finalizing rainfall intensity. The intensity of rainfall decreases with duration. To arrive at a fair estimate of intensity—duration for given frequencies, it is required to analyze the observed data on intensity duration rainfall of past records over a period of years in the area of interest. The longer the record available, the more dependable is the forecast. In Indian conditions, intensity of rainfall adopted in design is usually in the range of 12–20 mm/h.

Consultants have collected rainfall data for last 20 years from the meteorological department and the same has been adopted for rainfall analysis. The rainfall intensity has been worked out for storm frequency of once in 2 years—based on 20 years rainfall data 1986–2005. (Source: IMD's Cyclone Warning Centre). From the data the record rainfall of 270.50 mm at Visakhapatnam is of vital importance for this purpose. However, a rainfall of 320 mm, being 50 years—24 h maximum rainfall was considered for assessing the maximum flood discharge. As per the report (RBF No. 16) which was jointly prepared by Central Water Commission (CWC), IMD and Indian Railways, Visakhapatnam comes under Zone 4 (a) for which 50 years—24 h maximum rainfall is 320 mm. To plan and execute any hydrological projects and eliminate/reduce possibility of large-scale flooding over the affected area, it is essential to work out a reasonable rainfall input based on detailed study of rainfall pattern over and around this region. Since the catchment area involved is around 6.00 km^2 an input of 24-h or shorter duration rainfall is required for detailed study.

3.3 Specified Return Periods

Fifty years of 24-h rainfall for zone IV in which the present site is located is taken from Fig. 3.15(a) of RBF No. 16. However, hydrologists, particularly those concerned with small basins, as in present case, need estimates of rainfall for durations smaller than 24 h (which usually equals the time of concentration at the project site). Rainfall estimates corresponding to 6, 9, 12, and 18 h durations for various return period, worked out using the methodology developed by IMD and given in

CWC's publication titled "Flood Estimation Report for Eastern Coast Region—Sub zones 4(a), 4(b) and 4 (c)" are presented below.

Areal Reduction Factors The estimates of 24-h rainfall of shorter durations pertaining to a fixed location (viz. Visakhapatnam) and are termed as Point Rainfall estimates. However, the hydrologists need estimates of areal rainfall estimates corresponding to the areal coverage of the hydrologic project. In general the areal rainfall estimate is less than the point rainfall. Thus to obtain areal estimates of rainfall, certain reduction factors have to be applied to the point rainfall estimates. To achieve this, areal reduction factors (*Source: "Flood Estimation Report for eastern coast Region – Subzones 4(a), 4 (b) and 4(c)".—RBF No. 16*) corresponding to standard areas, ranging from 50 to 250 km^2.

3.4 Design Rainfall Input—The Basic Approach

The design of storm water drainage network would be based on the analysis of rainfall pattern and its distribution with respect to time in the catchment area. The critical parameters that would be considered with respect to the rainfall include frequency distribution, depth and duration. The data regarding the annual normal rainfall and annual coefficient of variation of rainfall would also be analyzed to generate rainfall features in the Corporation area. The frequency of Storm for which sewers are designed depends on the importance of the respective area to be drained. Commercial and industrial areas have to be subjected to less frequent flooding. The suggested frequency of flooding is Twice a year for Residential areas, once a year for Peripheral areas and Central and comparatively high priced areas and once in two years for Commercial and high priced areas. Sheelanagar is high priced and has industrial and commercial centers. The basic rainfall input for design of a hydraulic structure is an estimate of areal rainfall in a specified time interval. It depends on the size of catchments and time of concentration at the project site, along with its temporal distribution.

4 Hydrological Studies

4.1 General

The entire Sheelanagar and its surrounding areas were inundated during cyclonic heavy rainfall November and December 2010. This inundation had affected the normal life of the public and it is the fundamental duty of the civic body to address the problem, especially during cyclonic period. The flood is mainly caused by surplus water from Krishnamraju tank and due to non-availability of a defined surplus course which was existing prior to the rapid urbanization in that area.

The surplus course was modified in some stretches to suit the requirements of local people and this was the main root cause for all the havoc caused during the recent floods.

4.2 Design Flood for Surplus Channel from Krishnamraju Tank

Proper selection of the design flood is of utmost importance as this affects both the safety and the cost of the structure as well as for the residential and commercial buildings around. Too small design flood makes the design inadequate and thereby causing inundation on the downstream side. Hence the flood estimate has to be made accurately. The catchment area of Krishnamraju tank works out to 5.6475 km^2. The catchment area below Krishnamraju tank and up to Culvert/ National Highway (NH-5) has been calculated separately and it works out to be 0.3381 km^2. Hydro-meteorological approach (Unit Hydrograph) is adopted.

4.3 Characteristics of Gedda Flow

Krishnamraju tank is fed by a natural drainage channel basically of non-perennial nature and get the discharges mainly from precipitation coming through varying size of tiny rills and rivulets within the catchment. The hydraulic properties of this type of natural channels are irregular. In some contexts, empirical assumptions coupled with existing observations and experience is made so that the conditions of flow, in these channels become amiable to the analytical treatment of theoretical hydraulics. A through study of the behavior of flow in such natural channels requires knowledge of hydrology, geomorphology, sedimentation transport, etc. Existing bed slopes of the above natural drain, as it reveals, from the available longitudinal section that there is frequent change in bed slopes. The various studies carried out in connection with evolving flood control measures in the event of previous floods have been reviewed to understand the physical characteristics of the area and drainage system. The reports made available are given as follows.

4.4 Preliminary Design of Channels

Calculation of Discharge As discussed earlier, the preliminary designs of excavating a defined surplus water course of Krishnamraju tank is recommended as one of the flood control measures for inundation problem of Sheelanagar. The preliminary design of channel, earthen embankments, selection of side slopes, linings, if required, etc., are described in the subsequent sections. As a result of hydrological studies, surplus discharge of Krishnamraju tank works out to be 22.40 cumecs including the drainage discharge from the catchment area below Krishnamraju tank and up to NH-5. Therefore, it is required to design the surplus course so as to cater the total discharge of 22.40 cumecs preventing water logging of the area due flood drainage.

Selection of Channel Dimension: To cater the discharge, it is desirable to assign the dimensions of drainage channel on the basis of guidelines and recommendations so as to reach at economical and a practically feasible channel cross section.

Alternative 1 A Trapezoidal section of (3 m × 1.85 m) meters with a side slope of 1:1½ discharge.

Alternative 2 A Trapezoidal section of (3 m × 1.85 m) is with a side slope of 1:1½ to carry the required discharge. Four meter diameter pipe with partial flow has been proposed for the reach from 1.291–1.539 km. Circular conduit is proposed beyond NH-5 in view of the restricted land width available.

Alternative 3 A rectangular section of (5.85 × 1.85) meters is to be adopted to carry the required discharge.

Keeping in view the requirements of the GVMC, Alternative 3 is recommended. The drain provided will perform satisfactorily only if the following are addressed.

(1) The width of the existing culvert on NH-5 is 3.90 m in respect of right side lane and 4.50 m in respect of left side lane. This vent size is quite inadequate to pass a discharge of 22.40 cumecs for which the drainage channel was designed. Apart from this, the existing culvert is interfered with some pipe lines and thereby creating hurdles for the free flow of water. Hence efforts are to be made to modify the size of the culvert apart from removing the encroachments within the vent area.
(2) The size of the existing vent in GAIL compound wall is of size 6 m width. The size of existing drainage channel within the GAIL compound in continuation of the existing vent is (1.50 × 1.0) meters. With a bed fall of 1 in 800, the existing section is capable of carrying a discharge of about 3.58 cumecs as against a required discharge of 22.40 cumecs. Hence it is imperative to modify the size of existing vent in the GAIL compound wall as well as the existing section in continuation of this vent for free flow of water and to prevent the inundation problem on the other side of NH-5, i.e., towards Sheelanagar.
(3) The discharge after passing through the GAIL premises enters an existing tank near Panchavati colony of VUDA through an existing vent in the compound wall of GAIL. The size of the existing vent is of (3.50 × 1.20) meters. A velocity of 5.33 m per second is to be developed in order to carry a discharge of 22.40 cumecs which is highly improbable and unrealistic and as such it is imperative to modify the size of the existing vent on the GAIL compound for smooth flow of water as well as to prevent inundation of the GAIL premises.

(4) From the existing tank near Panchavati colony, the drain waters passes through a series of culverts existing on Port connectivity road before drains into the existing Naravagedda course on the downstream side of NH-5. The size of culverts are of size (2 m × 2 m) meters and are not in a position to carry the required discharge of 22.40 cumecs for which the drainage channel was designed.
(5) The existing drain courses and the existing tanks are filled with water hynith and other vegetation and in order to improve the carrying capacity, apart from modifying the section of drainage course, it is imperative to remove water hynith and other vegetation periodically so that there would be a smooth flow of water.
(6) The existing ground level/drain level near Yamaha show room, i.e., near Ayyappa temple is about +8.00 m and deep bed level of Naravagedda (Mehadrigedda surplus course) on the upstream side of NH-5 crossing is 3.494 meters. In order to provide quick relief to the inundated area, the aspect of providing a pilot channel from Yamaha show room to the existing Naravagedda course may have to be examined. An unlined section of 3 × 1.25 m is proposed. The Pilot channel is designed for 18% of the Theoretical discharge and is intended to provide temporary relief to the residents from submersion.

5 Conclusions and Recommendations

1. Flood mitigation plan for Sheelanagar has been evolved through ground Reconnaissance survey, Hydro-meteorological and Hydrological studies. Channel dimensions were obtained through uniform flow computation.
2. The surplus channel of Krishnamraju tank is designed for a max flood discharge of 22.40 cumecs. The bed level to be adopted at start is 6.00 and 4.07 mts at the end. It is proposed to connect the surplus channel to the existing vent available in the compound wall of GAIL yard on left side of NH-5. 3 alternatives were proposed.
3. The foreshore area of Krishnamraju tank is filled up with water hynith and other vegetation thereby reducing the capacity of the tank. Incidentally this situation has aggravated flood problem. It is recommended to remove water hynith and vegetation periodically from tank bed for restoring the capacity of the tank.
4. The sections of existing internal drains are kept as it is if they are able to carry the required discharge and proposed to be modified where ever it is found that the existing section is not able to carry the required discharge.
5. It is recommended to extend the existing drains at the places where they are discontinuous in order to have a connection to the main drain.

It is expected that above measures will give relief to the people of the affected area.

Acknowledgements The authors express their sincere thanks to The Commissioner, GVMC for granting this consultancy work.

Bibliography

1. Consultancy report on.: Flood Control Measures for Visakhapatnam Airport, by WAPCOS India Limited, November (2006)
2. Flood estimation methods for catchments less than 25 sq.km in area—RBF No. 16—prepared and published by Ministry of railways RDSO, Lucknow, Govt. of India (1990)
3. Improvement of Yerrigedda from Sankaramatham to FOB in port trust area—Reach—4 (from 3007.30 m to 3150 m)—DPR prepared by Aarvee Associates, Hyderabad (2005)
4. Khanna P.N.: Indian Practical Civil Engineers Handbook. UBS Publishers Distributors, ISBN-10: 8174767479 (2012)
5. IS 10430 2000.: Criteria for Design of Lined Canals and Guidance for Selection of Type of Lining. Bureau of Indian Standards, New Delhi, India
6. IS 7112-(2002).: Criteria for Cross Sections for Unlined Canals in Alluvial Soils. Bureau of Indian Standards, New Delhi, India
7. Santhosh K.G.: Irrigation Engineering and Hydraulic Structures. Khanna publishers, August (2009)
8. SI sheet No. 65 o/2 and o/3
9. Rajeswara Rao, K., Markandeya Raju P.: Design of Comprehensive storm water drainage system for Inundation problem in Sheelanagar area of Gajuwaka (Zone V), A DPR prepared for GVMC Visakhapatnam by Consultancy Division, MVGR College of Engineering, VIZIANAGARAM, INDIA (2012)
10. Ven Te Chow.: Open Channel Hydraulics. Published by Mc Graw-Hill Book Co. (1959)
11. Syed Azizul Haq.: Harvesting Rainwater from Buildings. Published by Springer International Publishing, ISBN 978-3-319-46360-5, 266 (2017)

Estimation of Reservoir Storage Using Artificial Neural Network (ANN)

P. Satish and H. Ramesh

Abstract The rapid growth in population increases water demand thus resulting in scarcity of water which is due to improper management rather than lack of resources. Reservoir is the most important source for surface water. So, reservoir storage plays a crucial role in efficient reservoir management. Artificial neural network (ANN) is capable of simulating reservoir storage capacity. So, in the present work five different feed forward back propagation ANN models by varying number of hidden layer neurons were developed for estimation of Harangi reservoir storage, Karnataka, India. The first 2 years (2010–12) data was used for supervised training and remaining data (2013–14) was used in prediction. The predictive accuracy using the statistical parameters like correlation coefficient (R) and mean absolute percentage error (MAPE) were found within the acceptable limit. Result shows that, ANN model with five hidden neurons (i.e., network architecture of 6-5-1) is performing well compared to all other models for prediction of reservoir storage estimation.

Keywords Artificial neural network · Reservoir · Feed forward back propagation
Prediction · Storage

1 Introduction

Extreme spatial and temporal variation of rainfall results in availability of surface water unpredictable. To cope this, reservoirs are constructed to store the water for future purpose and here comes the necessity of estimation of that storage for many reasons like reservoir operating policy, to know the capacity of reservoir, flood

P. Satish (✉) · H. Ramesh
Applied Mechanics & Hydraulics, National Institute
of Technology Karnataka, Surathkal, Mangaluru, India
e-mail: satishiiit07@gmail.com

H. Ramesh
e-mail: ramesh.hgowda@gmail.com

M. Rathinasamy et al. (eds.), *Water Resources and Environmental Engineering I*,
https://doi.org/10.1007/978-981-13-2044-6_5

control and many more. This needs real-time physical parameters on continuous basis. Artificial neural network is a computational model which works same as brain and having structure similar to biological neural network. It has a capability of correlating the input and output of any system.

Better understanding of input and output variables from statistical analysis is required prior to network modeling which results in coherent network design. A proper training-based neural network model is able to acquire the physical relation between the variables and may generate better results than conventional prediction techniques. Suitable ANN modeling is always advantageous in complex systems when compared with conventional modeling techniques [1]. Artificial neural networks have been accepted as a potentially useful tool for modeling various linear and complex nonlinear systems. In the hydrological forecasting and prediction context, ANNs have also proven to be a productive alternative to conventional methods for rainfall and stream flow forecasting, Groundwater modeling and reservoir operation [2]. Feed forward network having input, output and one hidden layer with sigmoid activation function to the monthly inflow data series and their performances were compared with an autoregressive integrated moving average time-series model. Jain et al. [3] concluded that the high flows are modeled better through neural networks. Optimal operating policies of a reservoir have been derived in deterministic and stochastic frame work as well as using ANN with five different network models by varying input values. Out of five the network which contains the current storage, current inflow and previous inflow as input variables and future storage as output (target) variable gives best result [4]. For important planning, design and management activities, we need accurate forecasted variables which affect the performance of the system. So, we need to choose suitable model for forecasting [5]. Forecasting of a river stage contains major input information for water resources systems planning and management. If such a forecasting is done on a continuous basis in the given year would help in providing a warning of the impending flood during high river stage and would assist in controlling reservoir outflows during the low river stage [6].

For this study, the most acceptable learning approach, i.e., supervised learning in training and feed forward back propagation ANN network model was used. Three years hydrologic data of Harangi reservoir was used. First two years (2010–12) of the data is used for training and remaining untrained available data (2013–14) is used to compare predicted data. Well accepted range of statistical parameters ensures the best model fit for this work.

2 Study Area

The focused area is Harangi reservoir. It is geographically lies between latitude 12°-29′-30″ North and longitude 75°-54′-20″ East. Near Hudgur village, Somwarpet taluk in Kodagu district in the Indian state of Karnataka. The reservoir is formed by a masonry dam built across the river Harangi. This is a major tributary

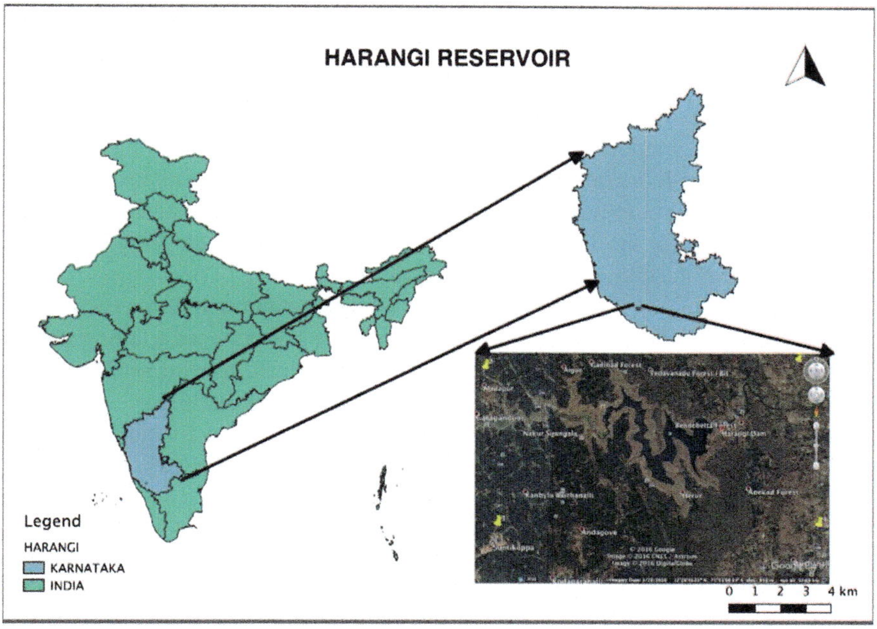

Fig. 1 Location of Harangi reservoir

to river Cauvery and has its origin in Pushpagiri hills, Coorg district situated on its western part and separated from Bhagamandala, the source of Cauvery. The Location of Harangi reservoir is shown in below Fig. 1.

2.1 Data Used

Data used in this work, contains hydrologic data including initial storage, inflow, irrigation sluice, river sluice, spillway, and evaporation loss for three years (2010–2012, 2013–2014) on daily basis of the Harangi reservoir, Hudgur village, Somwarpet taluk, Kodagu district of Karnataka, India. Data for the study was collected from the Water resources department, Government of Karnataka.

3 Theory and Methodology

Artificial neural network is a computational model which works same as brain and having structure similar to biological neural network. It has a capability of correlating the input and output of any linear and complex nonlinear system. The purpose of the neural network is to unfold the given task similar to brain would, although several neural networks are more abstract. Neural networks are still several orders of magnitude less complex than the human brain and closer to the computing power of a worm.

In feed forward back propagation network Input vector will propagate from input layer to the output layer through hidden layers. The output of the network is then compared with the desired output, and an error value is calculated for each of the neurons in the output layer. The error values are then propagated backwards from output layer to the input layer. Using this error values the weights are subjected to change to minimize the loss function.

To calculate the loss function gradient, a known desired output for each input is necessary. So, it is therefore usually considered to be a supervised learning method. It is a generalization of the delta rule to multilayered feed forward networks, made possible by using the chain rule to iteratively compute gradients for each layer.

Tan-sigmoid and pure line functions have been used as activation function in artificial neural networks more often. Basic artificial neural network architecture is as shown below Fig. 2.

The methodology used in this work starts with the collection of the reservoir data and preprocessing of that data. Using toolbox called "nntool" in MATLAB was chosen for this work. Data was divided into three sets (i.e., Training, Testing & Validation). These data sets were imported to MATLAB and developed different multilayered feed forward back propagation networks by varying number of hidden neurons and Levenberg-Marquardt as a training algorithm which is highly recommended supervised algorithm with six inputs such as Initial storage, Inflow, Irrigation sluice, River sluice, Spillway and Evaporation loss, and final storage as single Output/Target. Supervised training needs better distinction in input and output parameters.

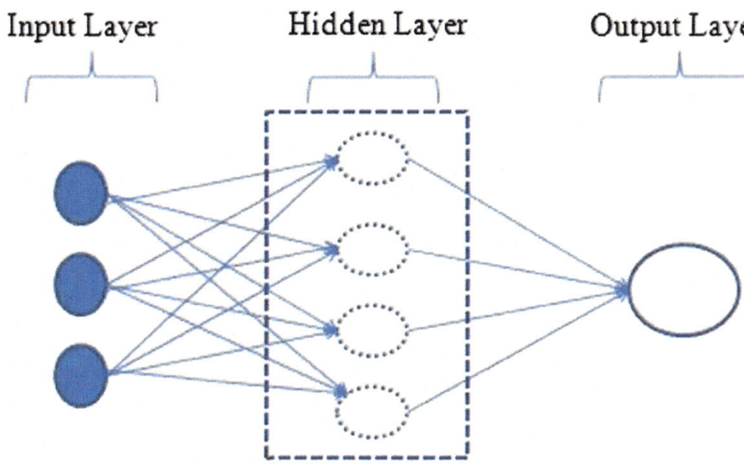

Fig. 2 Basic artificial neural network architecture

In this case it comes from the well established water balance Eq. (1) gave the clear understanding of relation between input and output parameters.

$$S_{t+1} = S_t + I_t - E_t - R_t, \tag{1}$$

where S_{t+1} = Future Storage; I_t = Current Inflow; S_t = Current Storage; E_t = Evaporation loss; t = Time increment; R_t = Outflow (i.e., irrigation, reservoir/power sluice and spillway).

Tan-sigmoid and pureline functions are used in input layer and hidden layer respectively as an activation function which helps in proper training of the network and gives the desired results. First 2 years (2010–12) data set used in training and remaining one year (2013–14) data set used in testing/prediction. Accuracy of network was found using the statistical parameters like correlation coefficient (R) and mean absolute percentage error (MAPE). Flowchart of methodology of artificial neural network was shown below Fig. 3.

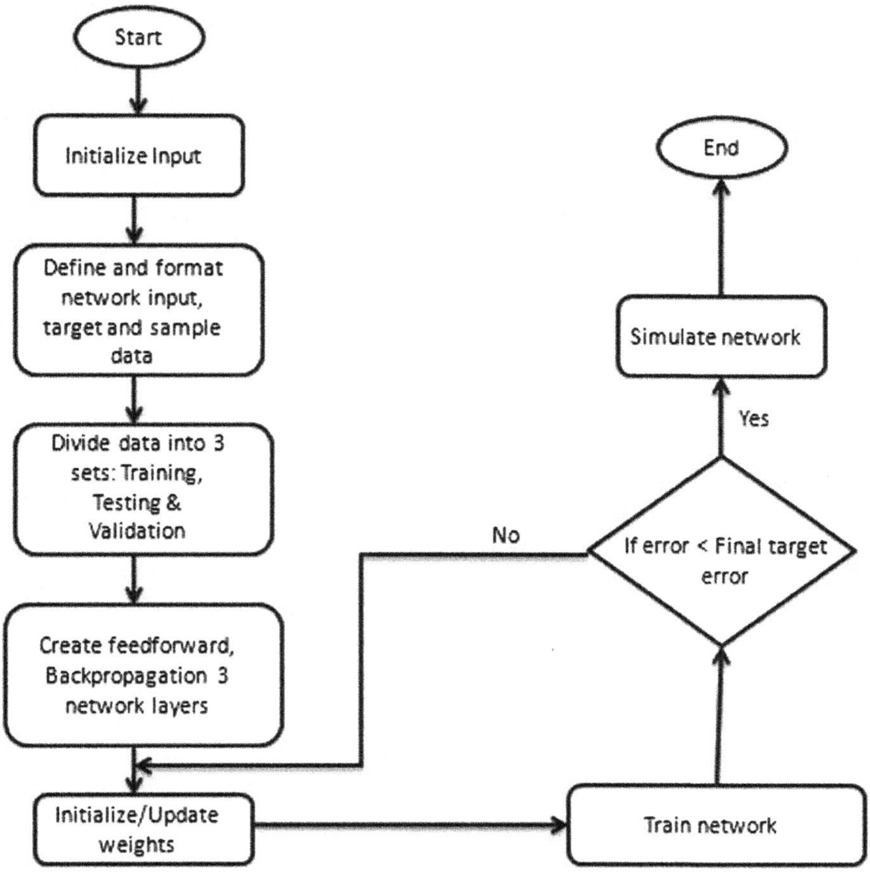

Fig. 3 Flowchart of methodology of artificial neural network

4 Results and Discussion

Five feed forward back propagation models with different network architecture are developed by toolbox in MATLAB named "nntool". The results were obtained by taking 2 years (2010–2012) of the data for training and the remaining data (2013–2014) on daily basis for prediction. The output results were checked for accuracy of prediction using statistical parameters like, correlation coefficient (*R*) and mean absolute percentage error (MAPE).

The analysis of all five networks in feed forward back propagation network (FFBN) was done after training, validation and testing. The results including some statistical parameters [i.e., correlation coefficient (*R*) and mean absolute percentage error (MAPE)] are presented in Table 1.

Based on the above results (Table 1), network with five hidden neurons of the feed forward back propagation performing good with correlation coefficient of 0.9882 and mean absolute percentage error (MAPE) of 5.225%. The observed and predicted storage of best network (6-5-1) is plotted as shown in Fig. 4.

Table 1 Network results including some statistical parameters

S. No.	Feed forward back propagation network (FFBN)		
	No. of hidden neurons	Correlation coefficient (*R*)	Mean absolute percentage error (MAPE) (%)
1	1	0.9778	13.629
2	2	0.9836	9.286
3	3	0.987	12.100
4	4	0.9828	5.618
5	**5**	**0.9882**	**5.225**

The bold values are implies that the network architecture corresponding to those values gave best results and suitable in this work

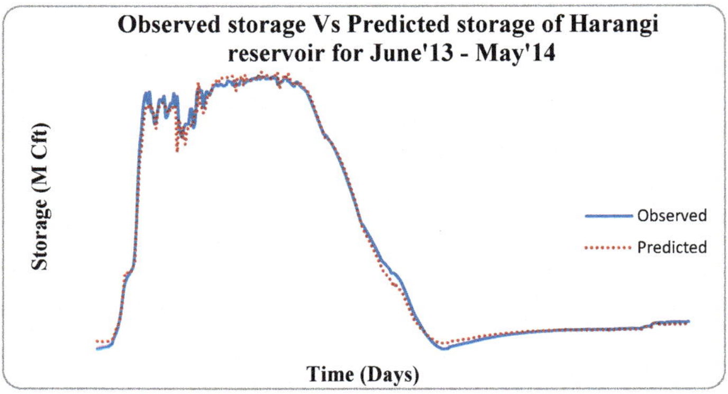

Fig. 4 Plot of observed and predicted storage (FFBN with 5 hidden neurons) of Harangi reservoir for June'13–May'14

From Fig. 4 it is observed that the predicted storage is nearly equal to observed storage throughout the year. Hence the feed forward back propagation network model with five hidden neurons is best suitable for present study.

Proper understanding of input and output parameters results in better training which leads to the good results/prediction.

The predicted storage will be useful for many purposes. Such as, to know the capacity of reservoir, reservoir operation policy and flood controlling and management studies.

Acknowledgements The authors are thankful to Water resources department, Karnataka for providing all the necessary data. Head of the department, all the teachers, family and friends who are helped us to complete this work.

References

1. Mustafa, M.R., Isa, M.H., Rezaur, R.B.: Artificial neural networks modeling in water resources engineering: infrastructure and applications. World Acad. Sci. Eng. Technol. **6**(2), 341–349 (2012)
2. Chang, F.J., Chang, Y.T.: Adaptive neuro-fuzzy inference system for prediction of water level in reservoir, Adv. Water Resour. **29**(1), 1–10 (2006). http://doi.org/10.1016/j.advwatres.2005.04.015
3. Jain, S.K., Das, A., Srivastava, D.K.: Application of ANN for reservoir inflow prediction and operation. J. Water Resour. Plann. Manage. **125**(5), 263–271 (1999)
4. Safayat Ali, S.: Optimal operation of single reservoir using artificial neural network. Int. J. Civ. Eng. Technol. **6**(6), 124–132 (2015)
5. Othman, F., Naseri, M., Reservoir inflow forecasting using artificial neural network. Int. J. Phys. Sci. **6**(3), 434–440 (2011). http://doi.org/10.5897/IJPS10.649
6. Thirumalaiah, Deo M.C.: River Stage forecasting using artificial neural networks. J. Hydrol. Eng. **3**(1), 26–32 (1998)

Sediment Distribution Pattern Studies for Thandava and Konam Reservoirs in Visakhapatnam District

J. Rangaiah, P. Udaya Bhaskar and V. Mallikarjuna

Abstract The life of any reservoir gets reduced due to deposition of sediment over a period. It is necessary in all reservoir operation plans to know the exact volume of water left after the sediment settles. Keeping this in view, two reservoirs in the Visakhapatnam district of state of Andhra Pradesh, viz; Thandava and Konam reservoirs have been selected. The main object of the present study is to identify the particular suitable theoretical method to predict sediment distribution pattern for selected catchments by comparing with the hydrographic survey results. Literature review has enabled the identification of two methods for determining the sediment distribution pattern. These methods are (a) Area-Increment Method (b) Empirical Area-Reduction Method. This exercise enables the identification of a suitable method for obtaining the sediment distribution in these reservoirs. Finally with the help of the method suitable for each of the selected reservoirs the current capacity (corresponding to year 2017) of the reservoirs is obtained.

Keywords Sediment distribution pattern · Area-increment method
Empirical area-reduction method

1 Introduction

1.1 General

A reservoir is an artificial lake and it begins to fill with sediment the moment it is formed and gradually dies from loss of its capacity caused by the deposition of

J. Rangaiah (✉)
Department of Civil Engineering, L.B.R.C.E, Mylavaram, Andhra Pradesh, India
e-mail: rangaiahj@gmail.com

P. U. Bhaskar
Department of Civil Engineering, JNTUK, Kakinada, Andhra Pradesh, India

V. Mallikarjuna
Department of Civil Engineering, V.R.S.E.C, Vijayawada, Andhra Pradesh, India

M. Rathinasamy et al. (eds.), *Water Resources and Environmental Engineering I*,
https://doi.org/10.1007/978-981-13-2044-6_6

sediment [1], most of which is transported by the inflowing water. One of the most important and intricate problems in the development and maintenance of storage reservoirs is the loss of storage capacity due to silting. Even through certain storage is reserved at planning stage for occupation by silt deposition, but the nature of sedimentation necessarily leads to silting [2, 3] and subsequent loss of a portion of live storage. Thus the sedimentation affects the capacity of the reservoir, which affects useful life of the reservoir in the longer run [4]. This problem needs careful analysis of the reservoir sedimentation for the continued functioning of the reservoir.

1.2 Need for Study of Reservoir Sedimentation

Sediment deposition in reservoirs is a complex and troublesome process. It piles up the debris behind a dam thereby reducing the capacity of the reservoir and its service function. Monitoring of the economic life of the reservoirs is undertaken by conducting sedimentation surveys at regular intervals because silting up of reservoirs is a continuous problem. In the view of the risk involved in silting, the reservoir sediment distribution pattern studies should be useful for periodically correcting the capacity curves to assure more efficient operation and also useful for determine the prevailing and probable future damages to a particular reservoir.

1.3 Scope of the Sediment Distribution Pattern Studies

Sediment distribution in a reservoir is very important and this aspect requires careful consideration in planning and design of the conservation structures. The designer is interested to know how high sediment will accumulate at the dam during a given period in order to fix up the sill elevation of the outlets and the penstock gate elevation. For the purposes of allocation of different storages, this is necessary in order to allow necessary storages for silt accumulation as silt distributed throughout the reservoir.

1.4 Identification of Problem

Sediment studies in reservoirs have become a vital filed of engineering study, due to the fact that reservoir sedimentation leads to the decrease of the water impounding capacity, reducing the useful life expectancy of the reservoir. Successive hydrographic surveys provide data for correcting the capacity curves and to operate the existing reservoir capacity efficiently [5]. But the conventional hydrographic survey methods are based on typical land survey techniques and are time-consuming for that reason, suitable theoretical methods are required for estimation of sediment accumulation.

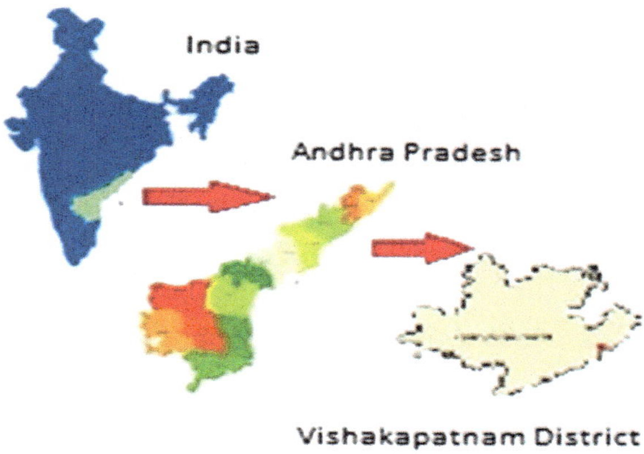

Fig. 1 Study area

1.5 *Study Area*

The Thandava and Konam Reservoir projects located in Visakhapatnam district are in the present investigation. Visakhapatnam district is a part of North Coastal Andhra Pradesh. The district is bounded on the north partly by the Orissa State and partly by Vizianagaram district, on the south by East Godavari district, on the west by Orissa state and on the east by Bay of Bengal.

The Thandava Reservoir project was constructed across river Thandava near Gantavani Kothagudem village in Kotauratla Mandal of Visakhapatnam district. The construction of the project was commenced in the year 1965 and completed in the year 1974 to create a storage capacity of 140.4286 M.cum at FRL. The global co-ordinates are, east longitude 82°-17′-20″ and north latitude 17°-17′-50″ respectively.

The Konam reservoir project [6] was constructed across the river Bodderu, a tributary to Sharada River at Konam village, in Cheedikadamandal of Visakhapatnam district. The Konam reservoir was constructed in the year 1979 to create a storage capacity of 24.0833 M.cum at FRL. The global co-ordinates are, east longitude of 82°-51′-50″ and north latitude 17°-58′-30″ respectively (Fig. 1).

2 Methodology and Data

2.1 *General*

The planning or design of a reservoir requires an analysis to determine how sediment deposits will be distributed in the reservoir. This is a difficult aspect of reservoir sedimentation because of the complex interaction between hydraulics of flow, reservoir operating policy, inflowing sediment load, and changes in the reservoir bed elevation. The traditional approach to analyzing the distribution of deposits has relied on empirical methods, all of which require a great deal of simplification from the actual physical problem. Borland and Miller of U.S.B.R., have suggested two methods in predicting the patterns. The first is strictly mathematical and is called "Area-Increment Method". The second method is a mathematical procedure based on the observations of sediment distribution in several reservoirs and is known as 'Empirical Area-Reduction Method'.

2.2 Area-Increment Method

This procedure is based on the assumption that the sediment deposition in a reservoir can be approximated by reducing the reservoir area at different elevations by a fixed amount. A series of approximations are involved in this.

The basic equation for this method is

$$V_s = V_o + A_o(H - h_o),$$

where

V_s = sediment volume to be distributed in the reservoir.

A_o = Area-correction factor in which is the original reservoir area at the new zero elevation at the dam.

V_o = sediment volume below new zero elevation.

H = Reservoir depth at the dam stream bed to full reservoir level.

h_o = depth to which reservoir is completely filled with sediment—new zero elevation.

The above equation mathematically states that the total volume of sediment V_s, consists of that which is uniformly distributed vertically over the height $(H - h_o)$ plus the portion below the new zero elevation of the reservoir.

According to the above, the following are the steps involved in computation:

1. h_o is obtained by trial and error by using the basic equation. The parameters V_s and H are the input and are derived from the data corresponding to the state for a particular year. The value of h_o is obtained with the help of the capacity area elevation curves so that the selected value of h_o satisfies the basic equation.
2. Here we find out the accumulative sediment volume by applying the area-correction factor at each depth increment.
3. In order to obtain the revised submergence area, the original area is reduced by the area-correction factor.
4. The revised capacity is arrived by subtracting the sediment volume from the original capacity at each increment.

2.3 Empirical Area-Reduction Method

From the sedimentation survey of several reservoirs in the United States, the distribution pattern of sediment has been classified in four standard types. The classification is as follows (Table 1).

The empirical area-reduction method for finding out the probable distribution of sediment involves in the following steps:

1. Classification of given reservoir into one of the standard types as already given above.
2. Computations by trial and error using the average end-area or primordial formula until the capacity computed equal the predetermined capacity.

Initially it is necessary to determine the Reservoir Type which fits a given reservoir. This is generally obtained by plotting the elevation-capacity on log–log paper, which gives a straight-line relationship and the slope of the line broadly indicates the type. Sometimes, the type may require suitable modification depending on the other characteristics of sediment and reservoir operation, etc. Finally, the adopted curve maybe a combination of two curves or may fall between two of the four curves, but generally it fits into one of the four basic types. Of course, considerable judgment is necessary in fitting the curve. These four standard curves (percent depth against per cent sediment) have again been converted into area-design curves for further use in the computation. The area-design curves have relative sediment area as the ordinate and relative depth as abscissa, the area under the curve in each case being equal to unity. The conversion from standard type curves to design-area curve has been accomplished by moody through equation $A_p = C\, P^m\, (1 - P)^n$, where A_p represents a dimensionless relative area at relative depth P above stream bed. C, m, & n are dimensionless constants which are determined by the type of reservoir. The characteristic constants, C, m, & n determined for the 4 types of the reservoir are as follows (Table 2):

Table 1 Classification of reservoirs for sediment distribution

M	Reservoirs type	Standard classification
1–1.5	Gorge	IV
1.5–2.5	Hill	III
2.5–3.5	Flood plain-foot hill	II
3.5–4.5	Lake	I

'*M*' is the reciprocal of the slope of the line obtained by plotting reservoir depth as ordinate and reservoir capacity as abscissa on log-log paper

Table 2 Dimensionless constants for equation $A_p = C\, P^m\, (1 - P)^n$

Type	*C*	*m*	*N*
I	3.4170	1.5	0.2
II	2.3240	0.5	0.4
III	15.882	1.1	2.3
IV	4.2324	0.1	2.5

The following steps are involved in computation.

1. Determination of the relative depth for each increment at the dam. This is merely the ratio of the incremental depth to the total depth.
2. Calculation of the values of relative sediment area (A_p) from the standard type curve selected.
3. Select a first approximation of the probable sediment elevation at the dam after sedimentation. Sediment area at each depth increment above the estimated new zero elevation is obtained by dividing the original area at zero elevation by the corresponding A_p value and multiplying this ratio (k) by A_p values at each succeeding increment.
4. With the sediment area established the incremental sediment volumes can be computed by the average end-area formula. If the sediment volume summation difference from that desired, next approximation is made.
5. The revised area is obtained by subtracting sediment area of satisfied approximation from original area at each increment.
6. The revised capacity is obtained by subtracting sediment volume of satisfied approximation from original capacity at each increment.

The sediment distributions obtained from the two methods (Area-Increment method and Empirical Area-Increment method) is compared with the hydrographic survey results.

2.4 Data

The elevation-area-capacity data of reservoirs are required for arriving at the sediment distribution and the same are collected from the original project report. Similarly the area and capacity corresponding to the stage of hydrographic survey is necessary for comparison purpose. This data is obtained from the records of APERL, Hyderabad. The data required for theoretical computations are presented here.

3 Discussion of Results

3.1 General

In the present investigation computations are carried out by using Area-Increment method and Empirical Area-Reduction method. The results from these two methods are compared with hydrographic survey data and are analyzed to draw conclusions.

3.2 Sediment Distribution from Area-Increment Method

The Area-Increment Method of determining the sediment is based on the assumption that the sediment deposition in a reservoir can be approximated by reducing the reservoir area at different elevations by a fixed amount. A series of approximations are involved in this. The reservoir capacities are calculated based on the reduced areas, applying either prismoidal formula or end-area method, such that the capacity below maximum normal level is the same as the predetermined capacity obtained by subtracting the sediment accumulation with time from the original capacity. The basic data required for the two reservoirs at different years. Following the methodology described in Sect. 2.2 the sediment distribution pattern corresponding to different elevations in the three reservoirs is arrived.

3.3 Sediment Distribution from Empirical Area-Reduction Method

In this method, sediment distribution pattern is considered to be dependent on the type of the reservoir which can be decided through a graph drawn between the depth and capacity for the original reservoir. From these graphs, it is seen that two reservoirs fall under the category of type II. The constants and coefficients to be selected in the sediment distribution equation depend on the type of reservoir as presented in Table 3. Considering one trial elevation to represent the zero level computations are carried out for obtaining the total sediment volume so that final value of V_s coincides with the original volume. The sediment distribution pattern for the reservoirs chose in the study is obtained from this method.

3.4 Comparison of Sediment Distribution Patterns from the Methods Considered

The results from the two methods are compared in the section that follows. Figures 2 and 3 shows the results pertaining to the elevation-area and elevation-capacity curves for Thandava reservoir. Figures 4 and 5 shows the results pertaining to the elevation-area and elevation-capacity curves for Konam reservoir.

Table 3 Basic data for sediment distribution computations

S. No.	Reservoir	Thandava	Konam	Units
1	Year of formation	1974	1979	
2	Year of sedimentation survey (hydrographic survey)	1991	1993	
3	Period of sedimentation	17	13	Years
4	F.R.L (from project report)	+115.825	+101.25	m
5	B.L (from project report)	+87.75	+82.00	m
6	Sill level (from project report)	+103.63	+87.00	m
7	Original depth at dam H = (F.R.L − B.L)	28.075	+19.25	m
8	Original capacity @ FRL from project report (year of formation)	140.4286	24.0833	M. cum
9	Total Capacity @ FRL from Hydrographic Survey	135.9862	22.3425	M. cum
10	Total sediment accumulated (vs.) from hydrographic survey	4.4424	1.7408	M. cum
11	Annual sediment inflow	0.2613	0.1339	M. cum

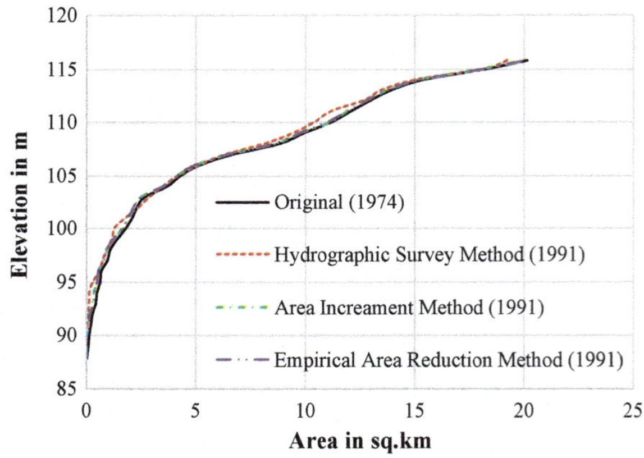

Fig. 2 Elevation versus area curves for Thandava reservoir (1991)

3.5 *Capacity Curves for the Current Year*

In the previous section it has been established that Area-Increment Method is suitable for Thandava reservoir and Empirical Area—Reduction Method is better suited for Konam reservoir. Figures 6, 7, 8 and 9 shows the elevation-area and elevation-capacity curves for the Thandava and Konam reservoirs for current year (2017) are presented here.

Fig. 3 Elevation versus capacity curves for Thandava reservoir (1991)

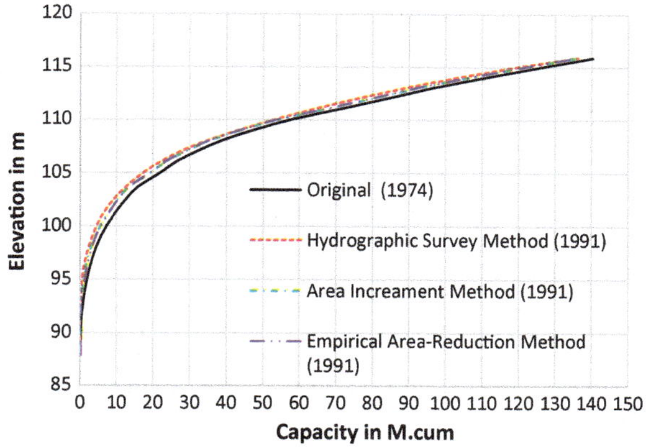

Fig. 4 Elevation versus area curves for Konam reservoir (1992)

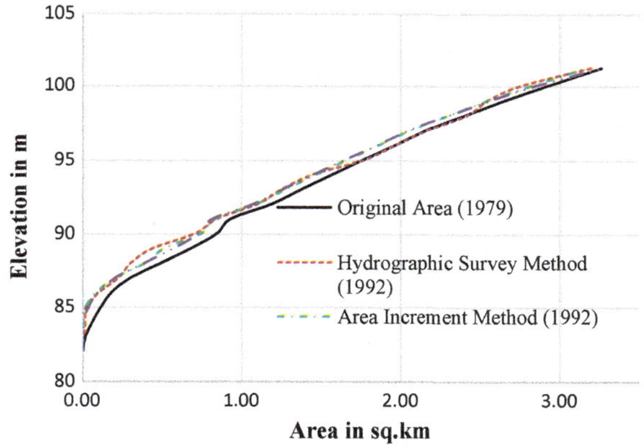

Fig. 5 Elevation versus capacity curves for Konam reservoir (1992)

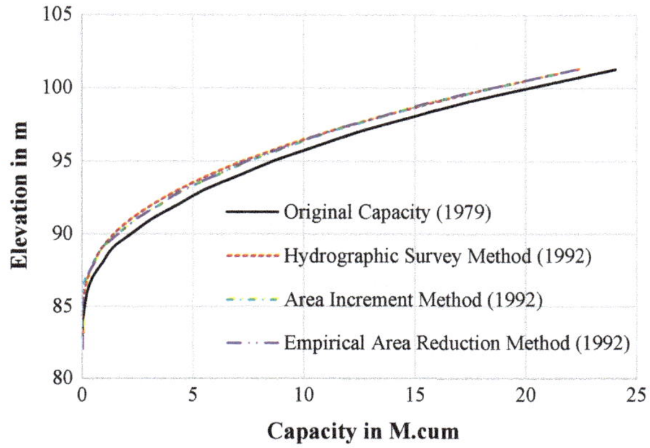

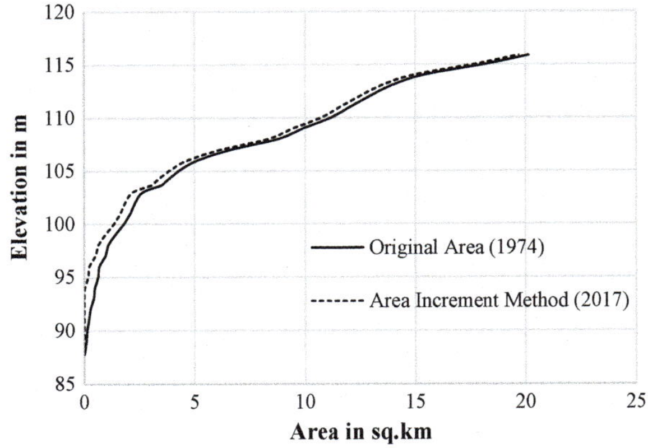

Fig. 6 Elevation versus area curves for Thandava reservoir (2017)

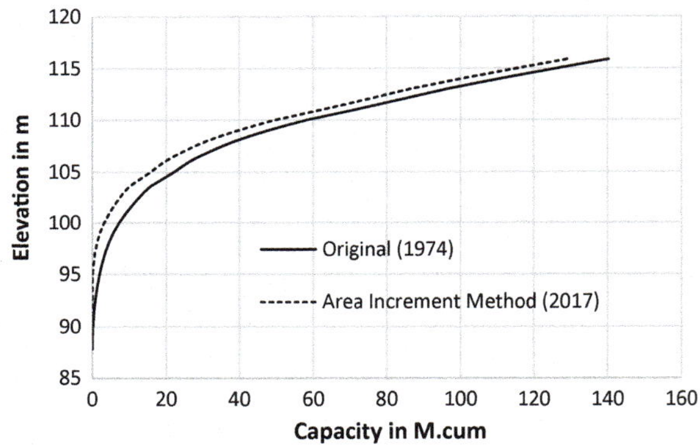

Fig. 7 Elevation versus capacity curves for Thandava reservoir (2017)

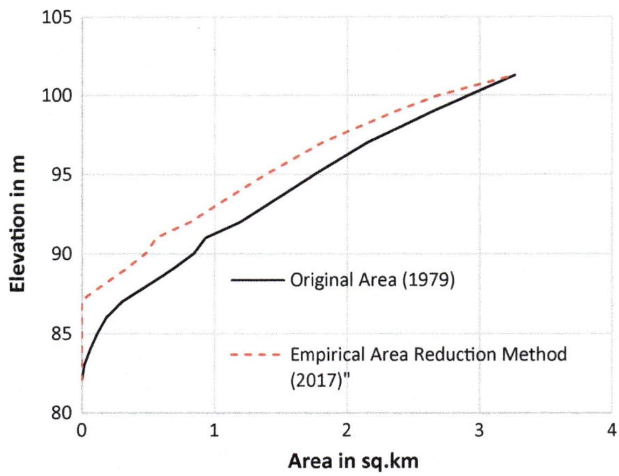

Fig. 8 Elevation versus area curves for Konam reservoir (2017)

Fig. 9 Elevation versus capacity curves for Konam reservoir (2017)

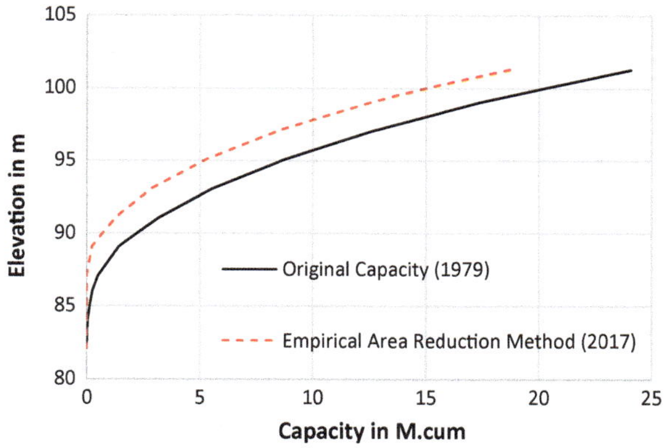

4 Conclusions

4.1 General

The objective of the present investigation is to identify a suitable equation for arriving at the sediment distribution in selected reservoirs. Finally with the help of the method suitable for each of the selected reservoirs the current capacity (corresponding to year 2017) of the reservoirs is obtained.

4.2 Conclusions

1. Two selected reservoirs fall under the categories of type II (Flood plain-foot hill).
2. The Area-increment method is predicting the capacity close to the observed capacity for Thandava reservoir.
3. The empirical area-reduction method is suitable for Konam reservoirs.
4. The reservoir elevation-area-capacity curves to the current year (2017) for the selected reservoirs are drawn.

References

1. Gottschalk, L.C.: Measurement of sedimentation in small Reservoirs, with discussions A.S.C.E. Transactions **117** (1952)
2. Joglekar, D.V.: Silting of reservoirs. J. Inst. Eng. (India) **40.8** (Part I) (1960)
3. Khosla, A.N.: Silting of Reservoirs. Central Board of Irrigation and Power. Publication No. 51 (1953)

4. Murthy, B.N.: Life of Reservoir. Publication 89, Central Board of Irrigation and Power, New Delhi (1980)
5. Report on the hydrographic surveys of Thandava reservoir, A.P. Engineering Research Laboratories, Hyderabad (1991)
6. Report on the hydrographic surveys of Konam reservoir, A.P. Engineering Research Laboratories, Hyderabad (1992)

Geospatial Data Requirements, Software, and Analysis for the Study of Floods in Urban Catchments

Ch. Ramesh Naidu

Abstract High intense rainfall due to rapid change in climate creates abnormal threat to engineers and city planners throughout the world and resulting urban areas being inundated from hours to days. This results in damage to infrastructure, public, and risk of epidemics. The recent floods in Mumbai and Chennai are some incitements to the public and government. Geospatial technology in conjunction with Remote Sensing plays an important role in flood modeling and mitigation analysis. Geospatial data serves as an input for hydraulic simulation and thereby the simulated results can further transferred back to GIS system for real world graphical representation of the potential flood vulnerability zones. There are many GIS, satellite data processing and hydraulic simulation tools/software is available in the industry. This article highlights the advantage of using a combination of open source software to arrive the objectives of flood hazard mapping and modeling.

Keywords Open source software · Storm water management · Remote sensing
Geographic information system · Digital elevation model (DEM)
Hydraulic simulation · Landuse/landcover

1 Introduction

A flood because of its intense nature generates severe loss of life and enormous damage to the building and properties. It occurs many places throughout the globe encompass almost 33% of the geographical area and to most of the population [1]. Analysis of floods involves study of hydrological data, hydraulic simulation of storm water drains both natural and manmade, terrain modeling and risk analysis. Hydrological analysis [2] involves collection of historical rainfall data from rain gauge stations, rainfall intensity, frequency and distribution analysis and calculation of flood volumes for which the corresponding hazards can be analyzed.

Ch. Ramesh Naidu (✉)
GVP College of Engineering, Visakhapatnam, Andhra Pradesh, India
e-mail: rameshnaidu@gvpce.ac.in

M. Rathinasamy et al. (eds.), *Water Resources and Environmental Engineering I*,
https://doi.org/10.1007/978-981-13-2044-6_7

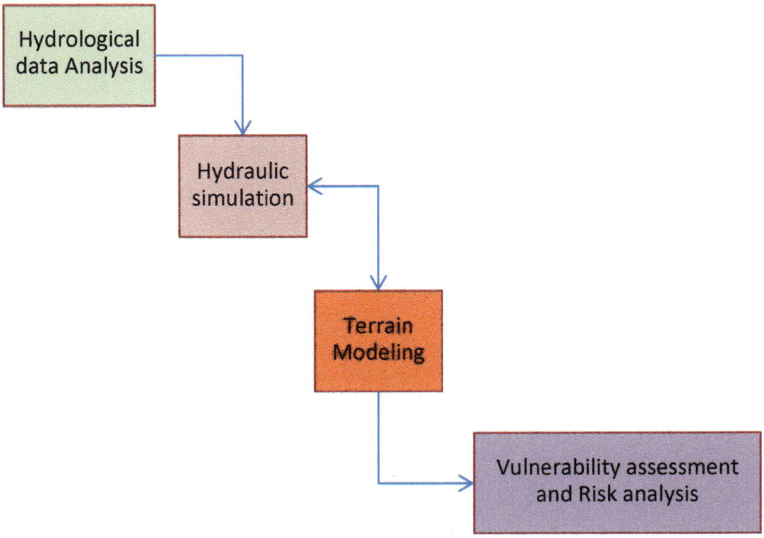

Fig. 1 Flow diagram (analysis of floods)

Flood volume is less in rural areas where rainfall infiltrates into the soil and vegetation. In contrast, developed areas, where built-up land runoff flows quickly to the drains and ditches. Uncontrolled urbanization has resulted in increase of surface imperviousness due to the rapid constructional activities in the form of buildings, roads and industrial activities. Due to this the storage capacity of natural drains and lakes reduces and thereby flooding takes place immediately.

Flood risk assessment and analysis comprises identification, risk and hazard assessment, risk analysis risk-vulnerability, and hazard estimation. The basic steps in the process are shown in Fig. 1.

Before the study of urban hydrology and hydraulic simulation, large-scale GIS data sets like climate, geology, soil, landuse/landover and topography needs to be developed for flood analysis. GIS tools are very useful and quite effective to provide solutions for managing and mitigating urban flooding [3]. These tools also assist in taking better, quick and alternate decisions during the disasters.

2 Methodology

The basic requirement for a researcher to develop the geospatial database is to have a combination of software which is customizable and freely obtainable to enable the users to access the available tools and resources. To cover this objective, there are three stages in the methodology and each stage requires a combination of software tools. These stages are

- Climate and rainfall analysis to determine the rainfall intensity and run off from the 1, 2, 3, 5, 10, 20 and 50 year return periods.
- Creation of geospatial database includes soil, landuse, DEM (Digital Elevation Model), property and infrastructure.
- Flood simulation and assessment to find out flood extent and overlay flooding results on property and administrative boundaries.

The data requirements are:

Climate and hydrology: Climate and meteorological data requirement to model and analyze intense rainfall and runoff processes in response to short time period. Due to the high-intense short-duration rainfall especially in hot weather climate, a well distributed spatially and temporally varied rainfall records are very essential.

Landuse/landcover includes buildings and critical infrastructure, Open spaces and forests: Landuse feature details with distribution, location and extent in GIS platform would be required for infiltration and run-off estimates. Different land use details should pertain to water bodies, forests, open spaces, residential, industrial, commercial, public/semi-public, etc., within the urban catchments [4].

Soil, geology, and geomorphology: Geological features and land formations in and around the settlements or towns would give brief idea about minerals and rock formations with bedding planes formed from the evolution of the earth and its development for example flood-effected- or flood-prone areas near to dams and natural river and stream courses. Geomorphology, geology, and lithology process also influence the type of structural, rock, and mineral compositions. This would result the assessment of quality and strength of the rocks by the engineers and planners in the construction activities. Soil type, texture, and composition will enable in the process of planning and designing the different urban assets like roads, buildings, bridges, water supply, storm water network, etc.

Topography, Slope, and aspect: Accurate ground surface, surface and sub surface drain elevation data including road center, road edge, drain top, and drain invert levels are important inputs for hydraulic simulation. LiDAR technology in conjunction with photogrammetry and DGPS facilitates the capture of terrain undulations and land surface features including detailed built-up and other landuse features. By Remote Sensing and Image processing software, satellite images and aerial photographs interpretation becomes easy, thereby minimizing costs compared to conventional surveys. High precision digital elevation models are required to generate slope, aspect and micro level catchments. Using surface hydrology GIS tools, flow accumulation and flow direction can be assessed and would help to earmark the catchments that undergo low, medium and high flood zones.

Natural drains and manmade drainage system: Flood analysis requires delineation of micro level urban catchments using the detailed topography with reference to natural and manmade drains. Study and capacity assessment of water bodies and natural drains are necessary for storage and diversion of excess floods.

Utility and Transportation network: Elevation, slope, and aspect maps in the GIS platform would help to take better decisions by integrating with buildings, open plots, utilities, and other important urban assets.

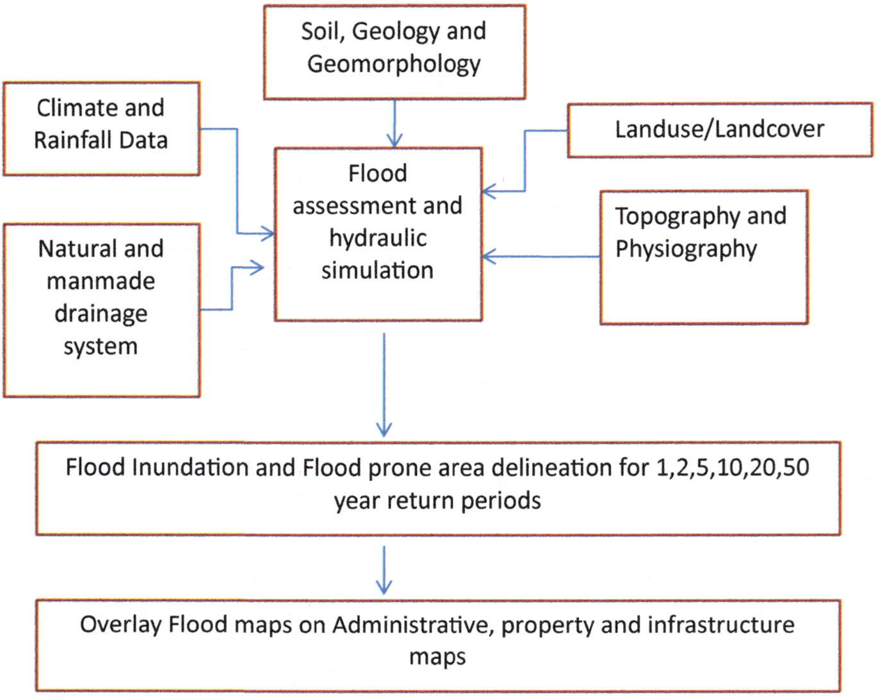

Fig. 2 Flow chart showing the process with key inputs

The proposed study is to assess the flood vulnerability in zone1 catchments of Greater Visakhapatnam Municipal Corporation area. The output from the process generates a series of maps which contains flood inundation scenarios for different return periods. Figure 2 shows the steps to analyze the flood vulnerability in GIS platform with reference to geospatial database. Table 1 elaborates the capability of the software tools that provides to process the required inputs and outputs to manage the study.

3 Conclusions

The study requires comprehensive geospatial data from recent high-resolution satellite imagery and field survey inputs and integration from the latest survey equipments. The tools in the software to be used for the proposed study are simple and easy to use and free of cost. The importance of long, continuous, reliable hydrological records is keys to estimate the minimum standard of protection. These kinds of studies may require improving, modernizing, and expanding existing network and rainfall and stream gauging stations to develop a comprehensive hydrological database. The hydrology and hydrographs of natural and manmade

Table 1 List of open source softwares to process various tasks involved in the study

S. No	Software	Type	Tasks
1	Quantum GIS (QGIS) [5]	Open source GIS software	• Digitization and development of thematic maps • Overlay and terrain analysis • Watershed and Viewshed analysis • 3D modeling
2	System for automated geoscientific analyses (SAGA)	Remote sensing and GIS data handling and digital image processing software	• Produces raster and vector maps • Produces DEM, Aspect and slope maps • Interprets landuse/ landcover maps from satellite images • Catchment delineation and watershed modeling • Calculation of overland flow distances to a river or channel network • Creates inundation maps for different time series
3	Environmental Protection Agency (EPA) storm water management model (SWMM) [6]	Storm water network and river channel modeling software	• Simulation of single and continuous rainfall events • Pipe capacity assessment • Supports dynamic flow routing • Modeling the flow in open channels and computing water surface profiles

drains should be modeled with urban micro-catchments so as to determine the necessary future developments with respect to increasing population. It is desirable to have an up to date landuse/landcover database in GIS platform that analyses the flood volume in urban catchments and to project the future threat of flooding due to the changes in the landuse. EPA SWMM is more suitable for urban catchments. Though there is no direct interface to utilize the GIS database, the data can be imported and exported with third-party solutions.

References

1. Dilley, M.: Natural Disaster Hotspots: A Global Risk Analysis, vol. 5. World Bank Publications, Washington, DC (2005)
2. Aksoy, H., Kirca, V.S.O., Burgan, H.I., Kellecioglu, D.: Hydrological and hydraulic models for determination of flood prone and flood inundation area (2016)
3. Guidelines on Management of Urban Flooding—National Disaster Management Authority, Govt. of. India
4. Elisabetta Genovese.: A methodological approach to land use-based flood damage assessment in urban areas. Prague case study (2006)
5. EPA's Storm Water Management Model (SWMM). https://www.epa.gov
6. Quantum GIS. http://www.qgis.org

A Review on Stability of Caisson Breakwater

Ajay Bhargav Gedda, Manu and Subba Rao

Abstract A Breakwater can be designed for several different purposes; the basic function of such kind breakwater is to protect the water region against waves. It provides a tranquility condition for ships to navigate, moor, and for cargo handling. These vertical structures (caisson) are more economical compared to the rubble mound breakwater, especially in deeper water depths. There is a demand to expand existing ports or to make them more profound, to provide a sufficient tranquil harbor basin in deep-water due to increasing draught of large vessels. The material required for rubble mound breakwaters increase quadratic with depth, but the volume of the caisson is less than that needed for a rubble mound breakwater because the latter increases with the square of water depth. This paper concerned the previous design and failure mechanism on caisson breakwater and highlights the future studies for such kind of breakwater.

Keywords Caisson breakwater · Tranquility · Failure mechanism

1 Introduction

Breakwater is constructed to maintain calm conditions in harbor basin or to maintain the stability of shoreline. Rubble mound breakwaters are of trapezoidal cross section and are built mainly of quarry stones and concrete blocks based on requirements. The selection of breakwater would be controlled by factors such as availability of rocks, depth of water, geomorphologic conditions of the seabed and the function of the breakwater. Caisson breakwaters are preferred in deeper depths as the quantities of quarried rocks increase significantly with increase in water depth which leads to increase in total project cost.

A. B. Gedda (✉) · Manu · S. Rao
Applied Mechanics and Hydraulics, National Institute of Technology Karnataka, Surathkal, Mangaluru 575025, India
e-mail: ajay.gedda@gmail.com

M. Rathinasamy et al. (eds.), *Water Resources and Environmental Engineering I*,
https://doi.org/10.1007/978-981-13-2044-6_8

When water depths are larger than 15–20 m, vertical breakwaters are preferred. Most of the existing vertical breakwaters were constructed using the massive pre-cast cellular reinforced concrete. Caissons are floated and then towed to the desired location. These are then ballasted and sunk on to the existing mound base and weight of which can be increased by rubble, concrete or sand fill. These types of breakwaters are also known as "Upright", "vertically composite" or "vertical faced or caisson breakwaters" [1].

The first vertical wall breakwater was constructed with a low rubble base in Naples in 1900. The vertical breakwater consisting of solid concrete blocks was first built Otaru port by I. Hirori, which lasted from 1897 to 1907 [2]. In rough seas, caisson breakwaters became the most suitable option. Wave forces and breakwater stability are the major factors which are governing the design methodologies of upright breakwater. The caissons were used as bulkheads and breakwater of many ports in Japan. Breaking waves exerts impulsive forces on an upright section of the caisson and those forces can be minimized by proper design and adoption of vertical breakwater covered with wave-dissipating blocks. Impulsive forces can also be reduced by providing a wave chamber and perforations on wave facing vertical wall. Caisson breakwaters with wave chambers are nowadays used for the base of oscillating water column. The Goda formula for design forces on vertical wall breakwaters accurately predicts forces under the most condition of wave breaking conditions [3].

2 Vertical Wall Breakwater Types

Functionally, mound breakwaters dissipate the wave energy of incident waves by forcing them to break on a slope and thus they do not produce much reflection. On the other hand upright or vertical wall breakwaters reflect the incident waves without dissipating much wave energy. Composite breakwaters function as mound breakwaters when the tidal level is low and as vertical breakwaters when the tidal level is high. According to functional point of view, vertical breakwaters are not limited to those erected directly on the sea bottom, but also include structures built on artificial foundation such as rubble mound.

A breakwater which consists of a caisson placed on a foundation with slopes on both sides and upright section of the breakwater is protected with armor block on the seaside called a composite breakwater. Figures 1 and 2 show the different types of composite breakwaters. These are used in locations where either the depth of water is large or the there is a large tidal range. In such situations, the quantity of the rubble stone required to construct a breakwater is very high and hence caisson breakwaters are preferred in that case.

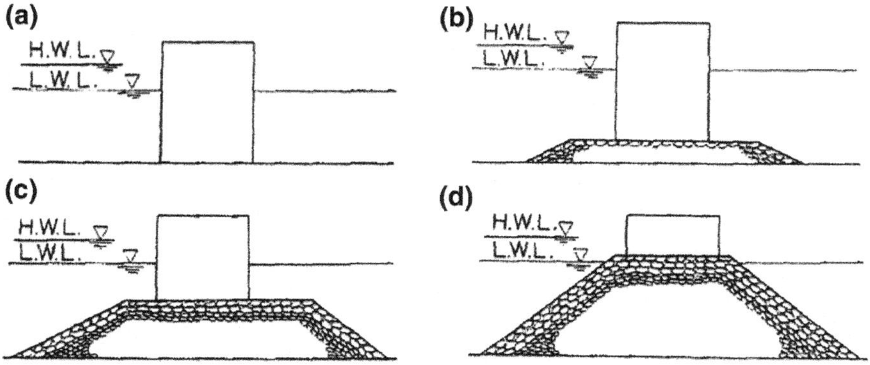

Fig. 1 Composite vertical wall breakwaters

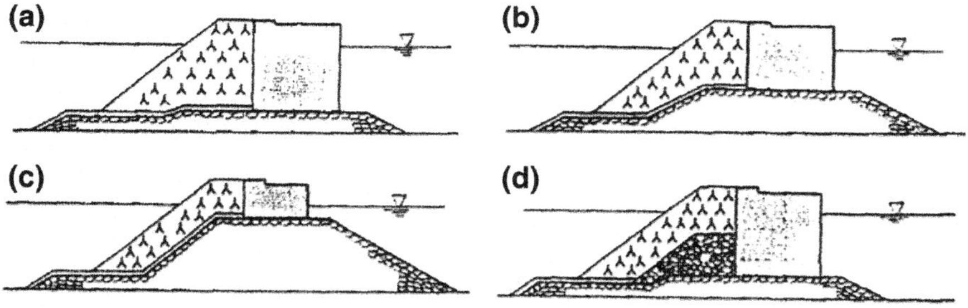

Fig. 2 Horizontally composite vertical wall breakwaters

3 Stability Analysis

Breakwater performance and stability depend upon the hydraulic parameters, i.e., Wave height, Wave period, Wave direction relative to breakwater direction, Duration of wave attack, Design water level. The stability analysis of vertical wall breakwater has to be performed against sliding; overturning and maximum soil stresses [4].

Apart from the above aspects, the vertical breakwater has to check for shear, liquefaction, etc. In stability calculation of breakwaters, the external forces such as the wave forces, hydrostatic pressure, buoyancy, and dead weight shall be considered.

Wave forces: For stability determination for a breakwater, the wave loads on all four sides have to be taken into consideration. Also, shock pressures have to account for are considered.

Hydrostatic pressure: Where there is a difference in the still water level between the harbor and seaward side of breakwaters, the hydrostatic pressure corresponding to the water level difference shall be considered.

Buoyancy: For the structural section of breakwater below the still water level, the buoyancy shall be considered. Dead weight: the dead weight of the breakwater may be calculated by using the unit weights of respective materials on the assumed section.

4 Design and Failure Mechanism of Caisson Breakwater

From the past studies, it is known that failure of caisson breakwaters is mainly due to sliding which is caused by impulsive wave forces acting on them. The three major reasons for failure are inherent to the structures itself, prevailing hydraulic and load conditions and the foundation and seabed morphology. The results suggest, the stability of vertical breakwater is a complex phenomenon, which can be solved by dynamic analysis and probabilistic design approaches. The disasters (such as sliding, failure of foundation etc.) occurred to vertical breakwaters are because of formation of clapotis due to wave reflection and impulsive forces exerted by the breaking waves [5].

Forces due to wave loads acting on vertical wall/caisson breakwaters arise from hydrostatic forces, hydrodynamic forces, uplift forces, geotechnical forces/reaction from supporting soil and internal water pressure. Geotechnical and wave forces are resisted mainly by the self-weight and friction between the supporting foundation [4].

Failure of composite breakwaters in Japan is occurring due to storm waves and the design of vertical wall breakwater should be designed to prevent and withstand total failure and subsequent damage due to heavy storms. Fully covered horizontal caisson breakwaters are more stable than ordinary composite breakwaters [6].

New spread parameter γ_{50} was introduced and which is the ratio of 50-year wave height to 10-year wave height to characterize extreme distribution functions. Design wave height and random wave breaking are key factors in performance-based design method [7].

Toe armor size, wave height, wave period, wave run-up, wave rundown and standing wave are the parameters influencing the stability of toe in front the vertical wall. Damage level increases with a decrease in toe armor size and increases in wave and wave period [8]. Placing an offshore structure on the seaward side of caisson/seawall can reduce the wave pressures on them. An increase in height of protecting offshore structure reduces the wave pressure on seawall/caisson. Irregular waves will be created even with regular wave attack if there is a presence of breakwater in front of caisson [9]. The results from the wave flume experiments on vertical wall breakwaters under regular and irregular wave attack are compared with Goda formula and found that Goda's theory is offering greater safety factor [10].

Kim and Suh [11] observed and studied the stability of caisson breakwater influenced by sea level rise and increase in wave heights due to climate change and found there was an insignificant effect of sea level rise on sliding of caisson breakwater when compared to the effect of an increase in wave height.

Esteban et al. [12] carried out laboratory experiments on caisson breakwater for sliding and tilting and observed the values for three storms of similar intensity. The authors also compared the experimental results with their computer model results and found the computer model studies are overestimating the deformation for first storm only and for remaining subsequent storms the values estimated are in the magnitude of the values obtained in a laboratory test. The authors concluded that this model can be used for estimation of final tilting and sliding.

Mase et al. [13] studied the displacement of wave-dissipating blocks, wave pressure and sliding of caisson type breakwater and related these three parameters. Authors found that sliding distance is more with blocks where repairing was not done than in the case, where the repairing was done and they also found if repairing is done after 5% damage, there will be a great decrease in caisson sliding distance.

Ruol et al. [14] measured uplift and horizontal wave pressures simultaneously with the movements of the caisson. He conducted experiments on different weights and geometries of caisson against both regular and irregular wave attack and found that existing formulae and experimental study gave nearly same results under breaking wave. Slotted vertical barriers are much cheaper than the rubble mound breakwaters as materials required for vertical slotted barriers is about 10% of material required for the rubble mound breakwater. Dynamic pressures acting on such barriers should be estimated accurately to make them more cost effective. Especially near still water, an increase in porosity leads to decrease in dynamic pressures [15].

5 Summary

In this paper, an attempt is made to summarize the previous work carried on caisson structures and the failure mechanisms of such structures. Nowadays the need of caisson type breakwater are increased due to the increasing draught of vessels, the flexibility of the structures drawn the attention to such kind of breakwater. Especially in deep-water compared to other kind of structures, the caisson structures are realized cost effective and saving the time constraints in construction. The stability of caisson type breakwaters is a complex phenomenon and it is dependent on its self-weight and frictional resistance between base of caisson breakwater and foundation, strength of foundation, permeability characteristics of foundation bed material. Changes in climatic condition give rise to the sea level which should also be considered for the design approaches. Oscillating water columns can be installed over the caisson breakwaters on seaward side. The abrupt collapse of structures is often progressive; by understanding the hydrodynamic behavior of structure, it could be prevented.

Acknowledgements The authors are thankful to the Director of National Institute of Technology Karnataka and Head of Dept. of Applied Mechanics and Hydraulics of the institute for providing access to wave flume facility and for their continuous support. Also, the authors gratefully acknowledge the financial assistance provided by Ministry of Earth Sciences under ESTC-COT cell, Govt. of India.

References

1. Franco, L.: Vertical breakwaters: the Italian experience. Coast. Eng. **22**(1–2), 31–55 (1994). https://doi.org/10.1016/0378-3839(94)90047-7
2. Tanimoto, K., Goda, Y.: Historic development of breakwater structures in the world. In: The Institution of Civil Engineers (eds.) Coastal Structures and Breakwaters. Thomas Telford, London, pp. 193–206 (1992)
3. Tanimoto, K., Takahashi, S.: Design and construction of caisson breakwaters—the Japanese experience. Coast. Eng. **22**(1–2), 57–77 (1994). https://doi.org/10.1016/0378-3839(94)90048-5
4. Allsop, N.W.H., Mckenna, J.E., Vicinanza, D., Whittaker, T.T.J.: New design methods for wave impact loadings on vertical breakwaters and seawalls. In: International Conference on Coastal Engineering, pp. 2508–2521 (1996)
5. Oumeraci, H.: Review and analysis of vertical breakwater failures—lessons learned. Coast. Eng. **22**(1–2), 3–29 (1994). https://doi.org/10.1016/0378-3839(94)90046-9
6. Takahashi, S., Shimosako, K.I., Katsutoshi, K., Kojiro, S.: Typical failures of composite breakwaters in Japan. Coastal Eng. 1899–1910 (2000)
7. Goda, Y.: Performance-based design of caisson breakwaters with a new approach to extreme wave statistics. Coastal Eng. J. **43**(4), 289–316 (2001). http://doi.org/10.1142/S0578563401000384
8. Deepak Dutt, M.: Experimental studies on the design of toe protection for vertical wall breakwaters. M. Tech thesis (NITK) (2001)
9. Munireddy, M.G., Neelamani, S.: Wave pressure reduction on vertical seawalls/caissons due to an offshore breakwater. Indian J. Marine Sci. **33**(4), 329–337 (2004)
10. Chiu, Y., Lin, J., Chang, S., Lin, Y., Chen, C.: An experimental study of wave forces on vertical breakwater. J. Mar. Sci. Appl. **15**(3), 158–170 (2007)
11. Kim, S.-W., Suh, K.-D.: Time-dependent performance-based design of caisson breakwater considering climate change impacts. Coastal Eng. **91**(1–11) (2012)
12. Esteban, M., Thao, N.D., Takagi, H., Shibayama, T., Link, L.S.: Laboratory experiments on the sliding failure of a caisson breakwater subjected to solitary wave attack, pp. 8–9, Nov 2008
13. Mase, H., Tsujio, D., Yasuda, T., Mori, N.: Stability analysis of composite breakwater with wave-dissipating blocks considering increase in sea levels, surges and waves due to climate change. Ocean Eng. **71**, 58–65 (2013). https://doi.org/10.1016/j.oceaneng.2012.12.037
14. Ruol, Piero, Martin, Paolo: Thomas Lykke-Andersen and Luca Martinelli.: Experimental investigation on caisson breakwater sliding. Coast. Eng. **2014**(94), 1–11 (2014)
15. Alkhalidi, M., Neelamani, S., Assad, A.I.A.H.: Wave pressures and forces on slotted vertical wave barriers. Ocean Eng. **108** (2015)

Application of Foam and Sand as Dual Media Filter for Rooftop Rainwater Harvesting System

Shilpa Mishra and A. R. Tembhurkar

Abstract This paper discussed the application of foam and sand as dual media filter for rooftop rainwater harvesting system for any residential or commercial building. The area selected for analysis is the new girl's hostel building of VNIT Nagpur campus. A rooftop rainwater harvesting system is proposed using foam and sand as dual media for filtration and various tests are performed and results are find to provide solution to water scarcity and sustainable use of water for future.

Keywords Foam and sand dual media filter · Rainwater harvesting Turbidity · Water conservation · Water reuse

1 Introduction

The area selected for research project is the new girl's hostel building of VNIT Nagpur campus. In VNIT, girl's hostel covered 4980 m^2 of land area. There are some single-seated-, double-seated-, and four-seated rooms provided for girls. Girls hostel consist of two buildings of one new and other is old one to accommodate around 200 girl students. New girl's hostel building has three floors and contains two sanitary blocks at each floor. The total sanitary facilities available in this building includes 12 wash basins, 18 bathrooms, 18 toilets, six squatting pad out of which every sanitary block contains two wash basins, three bathrooms, three W/C, one squatting pad and some spaces are provided for clothes washing on each floor of three floor building. Water is supplied to the girl's hostel through a corporation pipeline into a water sump of 36,000 L capacities which is pumped daily into overhead tank of 25,000 L capacities.

S. Mishra (✉)
Civil Engineering Department, MVSR Engineering College, Hyderabad, India
e-mail: shilpatiwarimishra0605@gmail.com

A. R. Tembhurkar
Department of Civil Engineering, VNIT Nagpur, Nagpur, India
e-mail: artembhurkar@civ.vnit.ac.in

M. Rathinasamy et al. (eds.), *Water Resources and Environmental Engineering I*,
https://doi.org/10.1007/978-981-13-2044-6_9

Based on the survey it is found that 62% of girls are satisfied and 38% are not satisfied with water supply and sanitation facilities in hostel because most of the time water is not pumped into the overhead tank. However, 48% of girls finding the good quality of water while 36% found moderate water quality due to which 86% of girls have not experienced any water-borne diseases. It is also found that 60% of girls are willing to pay extra charge for better water quality especially for drinking through small package water treatment plant and 40% do not want to pay. It should be noted that 95% of girls are agreed that water conservation is very important and recycling and reuse techniques such as grey water recycling for toilet flushing, gardening, etc., and rainwater harvesting should be adopted. To satisfy the need of water conservation which arises due to scarcity of water it is essential to adopt rainwater harvesting system in hostel premises as rain water harvesting system is made mandatory. Also, 85% girls are ready to use treated rain water. This also adds to the cost saving which makes it beneficial from management point of view also.

1.1 Quantification of Rain Water

A survey was conducted to collect baseline data for girl's hostel of VNIT Nagpur which includes population, water supply and sanitation facilities, water usage for individuals, etc. A seven-page questionnaire has been systematically prepared to generate data regarding the water usage pattern of the hostel inmates. Questions were also designed to know the current level of awareness and acceptance for rainwater harvesting in girl's hostel building. Survey was done in different years and different branch of girls.

The amount of water consumed was estimated based on the data obtained from the survey about the pattern of water usage and actually measuring the flows through various water fixtures is shown in Table 1.

Total quantity of rain water from rooftop is calculated from formula:

$$\text{Supply} = \text{Roof top coefficient} \times \text{Rainfall} \times \text{area of rooftop} \tag{1}$$

Table 1 Per capita consumption of water in a day

Activities	Per capita water consumption in L/day	Total (L/day)
Brushing	1.125	135
Hand washing	5.25	630
Face washing	5.5	660
Bathing	29.25	3510
Clothes washing	60.75	7290
Room cleaning	26.75	3210
Drinking	2.125	255
Toilet flushing	47	5640

Here,

Roof top coefficient = 0.9
Area of rooftop = 850 m^2
Cumulative supply of rainwater = 868.275 m^3/year.

1.2 *Characterization of Rain Water*

For deciding the treatment unit rainwater characterization was done as shown in Table 2. Rainwater sample was collected at first rain and after wash out. Laboratory experiments were conducted for the sample collected for finding different characteristics of rain water. All chemical analysis was carried out as described in Standard Methods [1]. Figure 1 shows the total water consumption in VNIT Nagpur girl hostel.

It can be seen from Table 2 the major pollutant which has to be reduced for purpose of brushing teeth, face wash and hand wash are turbidity, suspended solids, dissolved solids, total coliform.

Now the concept of foam and sand as dual media filter is applied for the treatment of rainwater harvesting and design of filter to fulfil the water requirement of girl hostel of VNIT Nagpur.

2 Proposed Rooftop Rainwater Harvesting System

Based on the above characteristics it was decided to provide a storage tank followed by dual media filter as a treatment unit. Since water required for brushing teeth, face washing, and hand washing should be free from turbidity, microbial contaminant and aesthetically good. It was found from experiments that the dual media filter

Table 2 Characterization of rain water

Characteristics	Range
Turbidity	1.1–8.9 NTU
Alkalinity	75–103.125 mg/l as $CaCO_3$
Chloride	0.08–20.1 mg/l
Hardness	24.0–44.24 mg/l
Dissolved oxygen	7.1–9.9 mg/l
Sulphate	0.36–0.48 mg/l
Total solids	48–60 mg/l
Dissolved solids	41–50 mg/l
Suspended solids	7–10 mg/l
pH	6.7–7.2

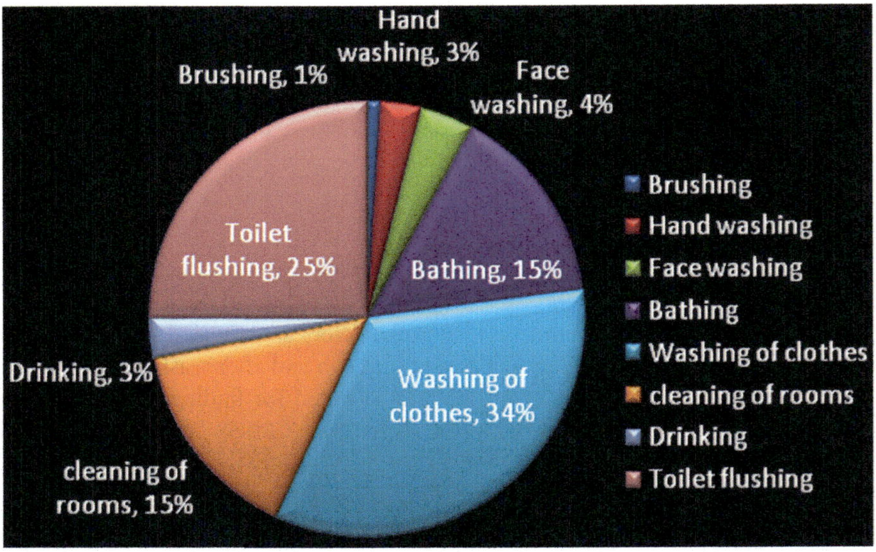

Fig. 1 Total water consumption in VNIT Nagpur girls hostel

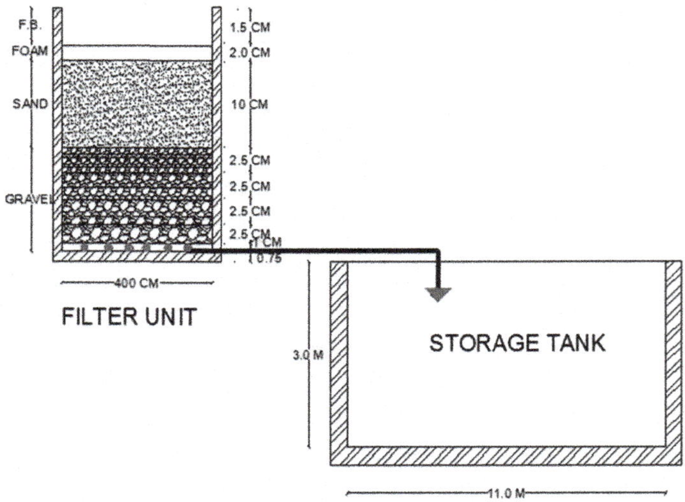

Fig. 2 Detailed drawing of unit proposed in girls hostel

having foam and sand as combination is suitable for treating rainwater [2–4]. A dose of disinfectant in the form of chlorine is advised to be administered during its utilization. Considering dual media filter as a treatment unit rainwater harvesting system was designed for hostel building as shown in Figs. 2 and 3 respectively.

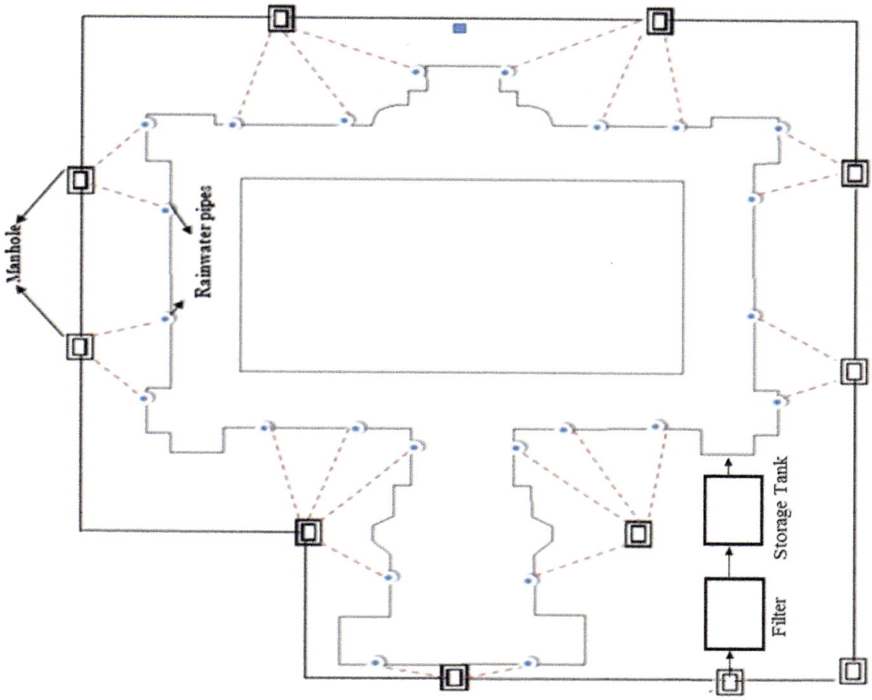

Fig. 3 Proposed design of rainwater harvesting system

Fig. 4 Comparison of various filters

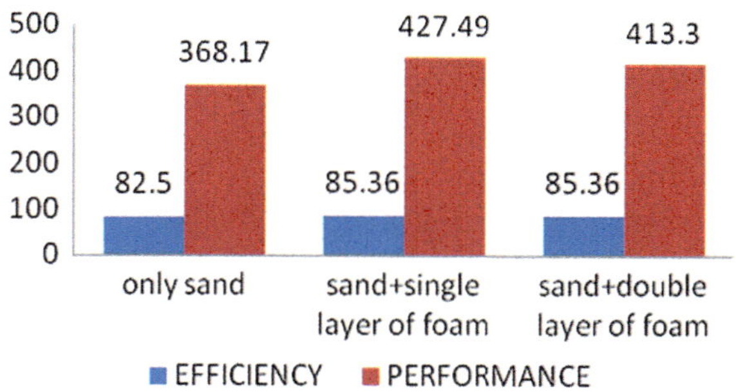

Figure 4 shows that the performance and efficiency comparison of various filters. As the performance and efficiency of sand plus single layer of foam is better than other filters, therefore we select sand plus single layer of foam as a filter for proposed design of rainwater harvesting system.

Table 3 Calculation of mass curve for roof top rainwater yield and water requirement by inmates of hostel

Month	Rainfall (mm)	No. of student	Yield (m^3)	Demand (m^3)	Cumulative yield (m^3)	Cumulative demand (m^3)	Deficit/ surplus
Jan	16	120	12.24	42.7	12.24	42.7	−30.46
Feb	2	120	1.53	39.9	13.77	82.6	−68.83
Mar	25	120	19.12	42.7	32.89	125.3	−92.41
April	20	120	15.3	42.7	48.19	168	−119.81
May	10	120	7.65	42.7	55.84	210.7	−154.86
June	174	30	133.11	10.68	188.95	221.38	−32.43
July	352	30	269.28	10.68	458.235	232.06	225.94
Aug	278	120	212.67	42.7	670.905	274.76	395.24
Sep	186	120	142.29	42.7	813.19	317.46	495.73
Oct	61	120	46.66	42.7	859.86	360.16	499.7
Nov	9	120	6.88	42.7	866.745	402.86	463.88
Dec	2	120	1.53	42.7	868.275	445.56	422.715

2.1 *Design of Storage Tank*

From the calculation of mass curve as shown in Table 3, maximum surplus and maximum deficit are 499.70 m^3 and 32.45 m^3 respectively.

Thus,

$$\begin{aligned} &\text{Total volume of water for which storage tank is to be designed} \\ &= 499.70\ \text{m}^3/\text{year} + 32.45\ \text{m}^3/\text{year} = 532.15\ \text{m}^3/\text{year} \end{aligned} \quad (2)$$

Assume Depth of tank = 3 m

Breadth to length ratio of tank = 1:1.5

Assume B: L = 1:1.5, B = 11 m, L = 16.5 m

$$\text{Area of tank} = \frac{\text{Volume}}{\text{Depth}} = \frac{532.15}{3} = 177.4\ \text{m}^2 \quad (3)$$

2.2 Design of Dual Media Filter

$$\begin{aligned} \text{Maximum Yield} &= \text{Maximum Intensity} \times \text{Roof top area} \times \text{Rainfall Coefficient} \\ &= 868.275\ \text{m}^3/\text{year} \\ &= 0.099\ \text{m}^3/\text{h} \\ \text{Rate of filtration} &= 0.526\ \text{m}^3/\text{m}^2/\text{h} \end{aligned} \tag{4}$$

$$\begin{aligned} \text{Area of filter required} &= \frac{0.099\ \text{m}^3/\text{h}}{0.526\ \text{m}^3/\text{m}^2/\text{h}} \\ &= 0.188\ \text{m}^2 \end{aligned} \tag{5}$$

Length = 0.4 m and Breadth = 0.4 m.

From the above qualitative analysis it is found that filtration forms are very vital treatment process for removing finer, suspended, colloidal, and microbial contamination [5–9]. Studies conducted for developing the dual media filter unit using sand at the bottom and foam at the top reveals the following:

- The dual media filter developed will be suitable for removing impurities having low turbidity such as roof top rain water harvest rain.
- The foam media at the top would facilitate easy cleaning of filter in case of rain water harvesting system as well as better removal of finer suspended solid impurity and microbial impurity.
- Installing this foam dual media filter is found to be technically attractive, operationally easy, feasible and low cost.
- The only problem faced by this filter is quick development of head loss which would require frequent wash however, since after first rain wash the quality of roof top rain water harvested contains less dissolved impurity and hence even though head loss development is quick but this can be utilized for sufficient operational time.
- The sand media used in this filter was with effective size from 0.35 to 0.70 and uniformity coefficient from 1.3 to 2.0 and tested under flow rate of 0.05–0.8 $m^3/m^2/h$ and varying turbidity from 0.5 to 8.0 NTU. It reveals that filter gives its best performance with sand media having effective size of 0.2 mm and uniformity coefficient of 2, flow rate 0.526 $m^3/m^2/h$ and influent turbidity of 8 NTU.
- The percentage removal for the sand filter is from 77 to 93% and for the best combination it is from 77.5 to 95.2%.
- It is also found that as the filter media effective size reduces and uniformity coefficient increases the quality of the filtrate improves however the volume of filtrate obtained reduces. The combined performance of the filter is judged by performance score which is found to be ranging from 134 to 427 during various batch studies. The operating conditions selected were mainly based on the

performance score giving maximum value as well as the efficiency. The maximum score obtained at optimum condition is 368. The studies on dual media (sand and foam) reveal that the sand depth of 10 cm and foam depth of 2 cm gives the best performance. The efficiency removal is found to be 95.12 as well as performance score is found to be 427. While assessing the application of this filter for treating roof top rain water harvesting, college girl hostel building was taken as a real-life field problem and attempt is made to provide design for this system.

On assessing the quantity and characteristics of roof top rain water from this building is found to be 868.2 m^3 roof top potential and has typical characteristics of pH = 6.7–7.2, Turbidity = 1.1–8.9 NTU, Total solids = 48–60 mg/l, Dissolved solids = 41–50 mg/l, Suspended solids = 7–10 mg/l, COD = 65–75 mg/l, Total Coliform = 4.5 MPN/100 ml, Faecal Coliform = 4.5 MPN/100 ml, *E. Coli* = 4.5/ 100 ml.

Results and analysis of rainwater harvesting system using proposed structure shows that the harvested rainwater can be recycle and reused for daily requirements [10].

3 Conclusion

Based on the survey it is also found that the hostel building has high potential for tapping the roof top rain water which can be utilized for partially satisfying the water demand. On analyzing the water demand and available quantity of harvested water it was found appropriate that the harvested rainwater can be used after treating through this foam and sand as dual media filter for brushing, face wash, and hand wash with little modification in existing pumping system. A dose of disinfectant in the form of chlorine is advised to be administered during its utilization. It can also be concluded that this foam and sand as dual media filter can be suitably utilized for treating roof top rain water harvesting system from buildings of institutional, residential type. This will not only supplement the existing water demand but will also be useful to provide solution to water scarcity and sustainable use of water for future.

References

1. APHA.: Standard Methods for the Examination of Waters and Wastewaters, 21st edn. American Public Health Association (APHA); Washington DC (2005)
2. Punmia, B.C., Ashok Kumar Jain, E., Jain, A.K.: Environmental Engineering—I: Water Supply Engineering, 2nd edn. Laxmi Publications (P) Ltd, India
3. Mouri, M., Niwa, C.: Pilot plant studies on filtration of raw sewage using floating filter media and multiple filter column inlets. Water Sci. Technol. **128**, 143–151 (1993)

4. Tiwari, S., Tembhurkar, A.R.: Laboratory studies on filtration unit with foam and sand as dual media. Int. Res. J. Lab Land **2**(8), 450–452 (2010)
5. Islamuddin, I.A.: Treatment of domestic waste water by filtration operation using low-cost natural adsorbents. Int. J. Emerg. Technol. Eng. Res. (IJETER) **4**(4), 114–118 (2016)
6. Wagener, C., Bellelo, S., Malone, R.: Static low density media filter for organic and solid removal from domestic wastewater. In: Proceedings of the 75th Annual Technical Exhibition and Conference of the Water Environment Federation, 2002, Chicago, Illinois (2002)
7. Badalians Gholikandi, G., Dehghanifard, E., Noori Sepehr, M., Torabian, A., Moalej, S., Dehnavi, A., Yari, A.R., Asgari, A.R.: Performance evaluation of different filter media in turbidity removal from water by application of modified qualitative indices. Iran J. Public Health **41**(4), 87–93 (2012)
8. Templeton, M.R., Andrews, R.C., Hofmann, R.: Removal of particle-associated bacteriophages by dual-media filtration at different filter cycle stages and impacts on subsequent UV disinfection. Water Res. **41**(11), 2393–406 (2007)
9. Devi, R., Alemayehu, E., Singh, V., Kumar, A., Mengistie, E.: Removal of fluoride, arsenic and coliform bacteria by modified homemade filter media from drinking water. Bioresour. Technol. **99**(7), 2269–2274 (2008)
10. Shegokar, V.V., Ramteke, D.S., Meshram, P.U.: Design and treatability studies of low cost grey water treatment with respect to recycle and reuse in rural areas. Int. J. Curr. Microbiol. Appl. Sci. **4**(8), 113–124 (2015)

Comparison of Saturated Hydraulic Conductivity Methods for Sandy Loam Soil with Different Land Uses

Aminul Islam, D. R. Mailapalli and Anuradha Behera

Abstract Saturated hydraulic conductivity (K_s) is a quantitative measure of saturated soil properties and it is essential for designing irrigation, drainage and waste water systems, modelling studies for understanding and prediting rates of infiltration, runoff, erosion, seepage, upflux, solute transport and migration of pollutant to groundwater. However, the accuracy of K_s is highly dependent on the method used, soil and surface characteristics. The objective of the study was to compare K_s methods such as two in situ [Double ring infiltrometer (DRI), air entry permeameter (AEP)] and one pedotransfer function (PTF) based methods for four different land uses such as paddy field (PADF), mango field (MANF), cashew field (CASF) and playground (PLAG). The K_s obtained from the DRI, AEP and PTF methods were used to study the effect of the method and land use on K_s and suitability of a method for a land use. It was observed that the measured K_s data using AEP and DRI of different land uses follow a log-normal distribution. The mean K_s were significantly different for both measuring technique and the land use. The AEP resulted highest (2.64 mm/h) and PTF lowest (1.59 mm/h) values of K_s, respectively for all land uses, whereas the K_s was highest (2.47 mm/h) and lowest (1.75 mm/h) for the land uses CASF and PLAG, respectively. For all land uses, the mean K_s were highest for AEP followed by DRI, and PTF methods. The order of K_s obtained for the land uses were CASF (2.51 mm/h), MANF (1.87 mm/h), PADF (1.82 mm/h) and PLAG (1.71 mm/h). Spatial variability of K_s was observed for DRI method and the land use PLAG. The selection of best suitable method for a particular situation can be obtained by optimizing the interdependent parameters, including method to be used, accuracy in instrument and measurement methods, soil condition and the numbers

A. Islam (✉)
Applied Engineering Department, Vignan's Foundation for Science, Technology and Research (VFSTR), Vadlamudi, Guntur 522213, India
e-mail: aminul.ubkv@gmail.com

D. R. Mailapalli · A. Behera
Agricultural and Food Engineering Department, Indian Institute of Technology, Kharagpur 721302, India

M. Rathinasamy et al. (eds.), *Water Resources and Environmental Engineering I*,
https://doi.org/10.1007/978-981-13-2044-6_10

of practical constraints of the investigation (e.g., cost, availability of manpower, time requirement, portability of estimate, simplicity in measuring technique, operating condition).

Keywords Saturated hydraulic conductivity · Ring infiltrometer Air entry permeameter · Pedotransfer function · Land use

1 Introduction

Saturated hydraulic conductivity (K_s) of soil is the ability to transmit water at saturated condition of soil [1, 2]. The saturated hydraulic conductivity (K_s) is helpful in designing irrigation, drainage and waste water systems, studying runoff characteristics and groundwater recharge, modelling of movement of water and pesticides from agricultural land [3–7]. Measurement of K_s at the top soil surface is at most important as most of the soil hydrological processes are influenced depth, land use, land cover and land management and enables development of hydrogeological models [8, 9]. Furthermore, the K_s is highly influenced by land and agricultural management, condition of the environment and measurement methods [10–13]. For accurate measurement of K_s, one should have proper knowledge on soil properties, selection of site, appropriate method and protocols are essential. The K_s methods have been classified into laboratory, in situ and correlation based methods [14, 15].

The laboratory methods include constant and falling head soil core methods, clod method and undisturbed soil block method [16–18]. The primary strengths of the laboratory methods include well defined and controlled water flow boundaries and fields; the flow environment can be specified and maintained [19]. During the soil sampling, artificial pores like cracks and fissures can be created inside the soil core and along the cylinder wall, which affect measurement of K_s [20]. The in situ methods include ring infiltrometer [21, 22], Guelph permeameter [23], double-tube method [24], disk permeameter [25] and air entry permeameter [26]. The in situ methods generally yield substantially different K_s values for the same location as they have different operating ranges, boundary conditions, underlying assassins [14, 27]. Air entry permeameter method cannot be applicable to wet soils as the wetting front determination will be difficult [26]. However, modified the air entry permeameter by adding a tension probe to detect wetting front easily [28]. Lee et al. [17] compared air entry permeameter, Guelph permeameter and falling head permeameter methods for different soil types and found that the K_s data were better described by log-normal frequency distribution and significantly different between some or all of the methods due to the presence of macrospores and air entrapment on each techniques. Mohanty et al. [29] compared disk permeameter, double-tube,

Guelph permeameter, velocity permeameter and laboratory methods at four different soil depths and reported that sample size and soil depth cause the K_s variability among the instruments. Bodhinayake and Si [30] measured infiltration rates with the help of double ring and tension infiltrometers for three different land uses and investigated that the K_s was 2–3 times higher for grasslands than that for cultivated lands because of high organic matter content. [31] Found higher K_s for tension infiltrometer while smallest K_s for soil core and intermediate for tension infiltrometer methods. Therefore, the accuracy of in situ measurement of K_s highly depends on the measurement method and soil condition [32–36]. The correlation methods include pedotransfer functions (PTF) based method [5, 37–41], which requires basic soil information such as soil physical properties (structure, texture, bulk density and organic matter content). The PTFs utilize these basic soil properties to perform mathematical formulations for estimation of K_s which is easier than other methods. But the estimation of these basic parameters is laborious, expensive and need more attention while measuring them in the field or laboratory. ROSETTA is a computer model developed based on neural network analyses include the bootstrap parameters with five hierarchical PTFs that estimate Van Genuchten moisture retention parameters and K_s using readily available soil information [42–44].

Spatial and temporal variability of K_s is generally influenced by porous media (macrospores, stones, fissures, cracks and root holes) and fluid media (viscosity and temperature), measurement methods and agricultural management [3, 13, 33, 45, 46]. Dev and Shukla [13] compared the K_s measured in the field and the laboratory and reported that the K_s was significantly different for the laboratory and field methods. Therefore, accurate measurement of K_s is difficult because it has spatial and seasonal variability and scaled dependency [47, 48]. It is important to know the particular soil and site conditions to select the proper application procedure and the method for getting reliable conductivity values [32, 49, 50]. Most of the K_s methods may not be accurate or appropriate for all soil conditions. Thus the selection of a suitable K_s method for a particular land use must consider some important factors such as initial cost, labour and time requirement, as well as accuracy [17]. Furthermore, there is no standard benchmark for measuring K_s and thus comparison of methods is essential source of information for choosing a suitable method for specific circumstances [14, 20]. In order to considering some limited studies focused on measuring K_s for different land uses and no study has been done (to the authors' knowledge) on comparison of air entry permeameter and ring infiltrometer, which share the same operating principle. Therefore, the objective was to study the performance of air entry permeameter and double ring infiltrometer methods for various landuse practices and compare them with pedotransfer function based methods.

2 Materials and Methods

2.1 Study Area

The study of field and laboratory experiments on K_s measurements were performed at Indian Institute Technology Kharagpur (IIT KGP) campus, which has an area of 2100 acres comprising agricultural, residential, academic, playground, parking and lawn areas. The climatic condition is humid subtropical with a mean annual rainfall of around 1700 mm out of which 80% of the total rainfall occurs in the month of June to September, i.e., in monsoon period. Winters are brief but chilly, lasting from December to mid-February (100–250 °C), summer (March to June) is hot (250–400 °C) and sometimes humid (50–95%). Four land uses such as paddy field (PADF) as an agricultural area, mango and cashew field as orchard area (MANF and CASF) and Tata Sports Complex as playground (PLAG) were selected for measuring K_s at 0–10 cm depth of top soil. The five years' past record of these land uses is presented in Table 1.

2.2 Soil Physical Properties

Soil core samples were collected from the top 10 cm depth from three locations of each land use. The soil samples were analyzed for determining % sand, % silt and % clay and bulk density using standard soil analysis protocols [51]. It was observed that the soils for all land uses were mostly of sandy loam type. The mean sand and silt contents were not much different among the land uses at 5% level of significance, but the clay content was significantly different, and highest for PADF with 18.33% and lowest for MANF with 14.00%. The highest clay content in PADF was mainly due to soil puddling and addition of organic carbon during cultural operations. The bulk density was not significantly different among the land uses and varied from 1.72 to 1.85 g/cm^3. Similarly, the porosity was not significantly

Table 1 Soil condition for different land uses of the study area

Land use	Land use history for past 5 years	Ground condition
PADF	*Kharif*: paddy and *Rabi*: wheat/paddy	Presence of cracks, paddy roots, stems and weeds
MANF	Not tilled	Completely shaded, and very light weeds between plants
CASF	Not tilled	Loose soil, with long weeds (4–12 cm), many worm holes
PLAG	Soil compacted annually	Covered with grass (3–5 cm), compact and mixed with gravels/plastics

Table 2 Descriptive statistics of selected land uses soil properties

Soil properties	Statistic	Land use type			
		PADF	MANF	CASF	PLAG
Sand percentage (%)	Mean (%)	64.33	68.67	69.67	63.33
	SD[a] (%)	3.51	1.15	8.96	1.15
	CV[b] (%)	5	2	1.3	2
	SE[c] (%)	2.03	0.67	5.17	0.67
	SV[d] (%)	12.33	1.33	80.33	1.33
	Range	7	2	16	2
Clay percentage (%)	Mean (%)	18.33	14.00	15.67	16.00
	SD[a] (%)	1.53	1	2.08	3.46
	CV[b] (%)	8	7	1.3	2.2
	SE[c] (%)	0.88	0.58	1.20	2.00
	SV[d] (%)	2.33	1	4.33	12
	Range	3	2	4	6
Bulk density (g cm^{-3})	Mean (g cm^{-3})	1.85	1.78	1.73	1.75
	SD[a] (g cm^{-3})	0.05	0.08	0.08	0.1
	CV[b] (%)	2	5	5	5
	SE[c] (%)	3	5	5	6
	SV[d] (g cm^{-3})	0	0.01	0.01	0.01
	Range	0.1	0.16	0.159	0.19
Porosity (%)	Mean (%)	33	33	35	34
	SD[a] (%)	2	3	3	4
	CV[b] (%)	6	9	9	1.1
	SE[c] (%)	1	2	2	2
	SV[d] (%)	0	0	0	0
	Range	4	4	6	7

[a]Standard deviation
[b]Coefficient of variation
[c]Standard error
[d]Sample variance

different for the land uses and varied from 33 to 35%. Table 2 presents the descriptive statistics of the soil physical properties of the different land uses of the study site.

2.3 Methods of K_s Measurement

The K_s was measured at three random locations for each land use using double ring infiltrometers and air entry permeameters. The pedotransfer function based method was used for estimating K_s by using the physical properties of the soil.

2.3.1 In Situ Measuring Techniques

Double Ring Infiltrometer

Double ring infiltrometer (DRI) device and Green and Ampt (1911) model were used together for estimation of K_s. The DRI consists of an inner ring of 30 cm diameter and an outer ring of 60 cm diameter and 2 cm thickness inserted into the ground at 10 cm depth at the study site with taking care the ring sides vertical and minimum soil disturbance. Water was subjected to fill the ring up to a depth of 10 cm to each ring which was marked by a point gauge to measure the reduced water level due to infiltration. A constant head of 10 cm water was supplied to each ring and the outer ring helps to control the lateral flow from the inner ring. The K_s can be estimated when water flow rate inside the inner ring comes to a steady state. The observations were used to generate data sets for infiltration rate and infiltration depth values with respect to different time intervals. The infiltration data obtained from the DRI method were fitted with the Green and Amp infiltration equation and K_s values were estimated by using Eqs. 1 and 2

The infiltration data were fitted with a physically based Green and Ampt (1911) infiltration model, to estimate K_s [52]. The Green and Ampt infiltration equation can be written as:

$$f = K_s\left(1 + \frac{\eta S_c}{F}\right) \tag{1}$$

where, F is the cumulative infiltration [L]; f is infiltration capacity [L T^{-1}]; η is porosity of soil; S_c is capillary suction at the wetting front and K_s is field saturated hydraulic conductivity [L T^{-1}].

Equation 1 could be considered as a straight line equation;

$$f = p + \frac{q}{F} \tag{2}$$

where, F is the cumulative infiltration [L T^{-1}]; f is the infiltration capacity [L T^{-1}] and p and q are the Green-Ampt parameters of infiltration model. The values of f are plotted against ($1/F$) and the best fit straight line was drawn through the plotted points. The intercept of the best fit line gives the K_s [52, 53].

Air Entry Permeameter

The operating mechanism of air entry permeameter (AEP) is almost equivalent to the single ring infiltrometer in design and operation. The vertical saturated hydraulic conductivity was measured in this method relating to volumetric flux of water entering into the soil. The major differences between ring infiltrometer and AEP methods are the instrument used in AEP method completely penetrates into the deeper layer of the soil which also measures the pressure of air inside the soil. The AEP consisted of a single permeameter ring of 30 cm in diameter and 25 cm-long, closed with an air-tight cover at the top, which was forced into soil depth up to 15–25 cm and infiltrates water through the cylinder under a positive pressure head about 1 m [17]. The pressure head was used for determination of hydraulic gradient and K_s measurement. Water was introduced into permeameter through stand pipe and allowed freely to infiltrate into the soil. The rate of infiltration was observed and continued up to a flow rate of relatively stable in nature. The air entry value is calculated when there is a minimum pressure was being measured over the standing water inside the permeameter ring.

$$P_a = G + L + P_{min} \tag{3}$$

where, P_a is the air entry value of soil, expressed as pressure head in cm at the point of entry are negative value [L]; P_{min} is the minimum pressure in cm water as determined by maximum reading on vacuum gauge are negative value [L]; G is the height of vacuum gauge [L]; L is the depth of wetting front [L].

$$K_S = \frac{2 \times (\mathrm{d}H/\mathrm{d}t) \times l \times (R_r/R_c)^2}{(H_t + L + P_a/2)} \tag{4}$$

where, $\mathrm{d}H/\mathrm{d}t$ is the rate of fall of water level in reservoir just before closing supply valve [L T^{-1}]; H_t is the height above soil surface of water level in reservoir at time supply valve is closed [L]; R_r is the radius of reservoir [L]; R_c is the radius of cylinder [L].

2.3.2 Pedotransfer Function Based Method

The pedotransfer functions (PTF) are the predictive function of soil physical properties using data from soil survey. ROSETTA (R) implements five hierarchical PTFs conceptualized on neural network analysis combined with the bootstrap method, and it allows predicting uncertainty in K_s [43, 54]. The input parameters for the K_s estimation such as textural classes, silt, and clay percentage in fraction, soil dry bulk density and soil volumetric water content and water suction were used at five different levels [55]. ROSETTA (R) relates between volumetric water content (θ) and water suction at a head of (h) which is known as water retention [$\theta(h)$] as well as saturated hydraulic conductivity which can be better described by Mualem–Van Genuchten equation [56] given as

$$\theta(h) = \theta_r + \frac{\theta_s - \theta_r}{[1 + (\alpha \times h)^n]^m} \tag{5}$$

where, $\theta(h)$ is soil volumetric water content at suction head h [L^3 L^{-3}]; θ_s and θ_r = saturated and residual water content [L^3 L^{-3}]; α is the inverse air entry suction; n is a measure of the pore-size distribution and $m = 1 - 1/n$.

The hydraulic conductivity can be described by Mualem-Van Genuchten model in conjunction with pore-size distribution model as:

$$K(S_e) = K_0 S_e^L \left[1 - \left(1 - S_e^{\frac{n}{n-1}}\right)^m\right]^2 \quad (6)$$

where, K_0 is a fitted value of K at saturation [L T^{-1}], which is similar but not considered as K_s and L is a pore connectivity factor and S_e is the effective saturation given as

$$S_e = \frac{(\theta(h) - \theta_r)}{(\theta_s - \theta_r)} = \left[\frac{1}{1 + (\alpha \times h)^n}\right]^m \quad (7)$$

2.4 *Analysis of* K$_s$ *Data*

The K_s data measured or estimated by three methods (DRI, AEP and PTF) from four land uses (PADF, MANF, CASF and PLAG) were pooled together and divided into three groups; (1) K_s data-method (4 land uses $\times$ 3 trials = 12 data points for each method), (2) K_s data-land use (3 methods $\times$ 3 trials = 9 data points for each land use) and (3) K_s data-total (3 methods $\times$ 3 trials $\times$ 4 land use = 36 data points). These data groups were analyzed for normality test and best fit frequency distribution by using the Shapiro and Wilk (1965) test at 5% significant level. Based on the best fit frequency distribution, the K_s obtained from the DRI, AEP and PTF methods were compared relatively based on descriptive statistics, analysis of variance (ANOVA) and Tukey's pair-wise mean comparison test.

3 Results and Discussion

3.1 *Characteristics of* K$_s$ *Data*

The nature of K_s data for AEP method rejected the null hypothesis (data is normally distributed) at 5% significant level ($P < 0.05$) and hence, the K_s data (K_s data-methods) for these methods were approximately not normally distributed. This is evident from the skewness and kurtosis values, which are not close to zero. This is also evident from the normal *Q-Q* plot of AEP given in Fig. 1a (the plots for other methods are not shown here), in which the data points do not lie on a straight line. The null hypothesis was not rejected for the K_s data of DRI and PTF and thus the K_s data for these methods were approximately normally distributed. With the

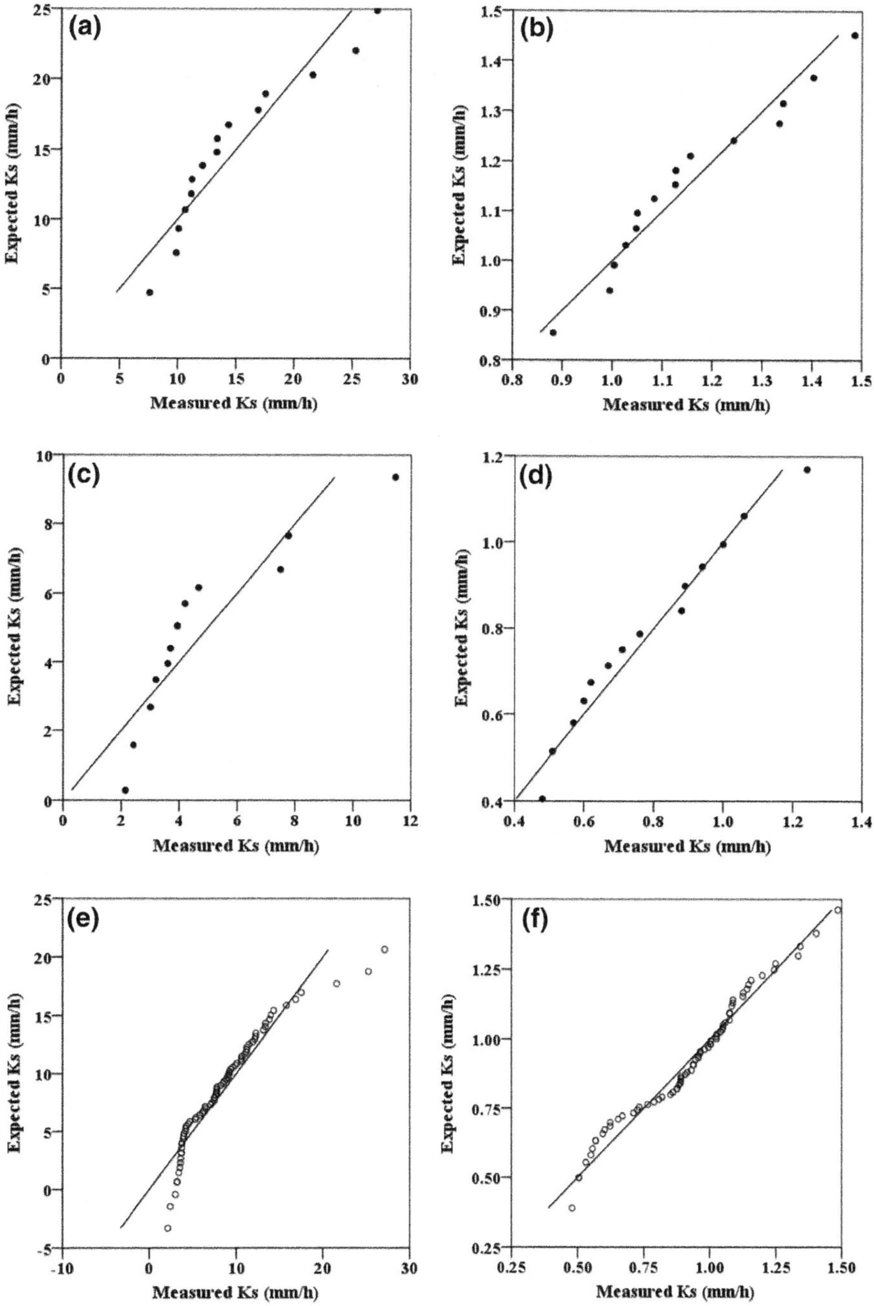

Fig. 1 Normality test results using *Q-Q* plots for **a**, **b** K_s data-AEP method, **c**, **d** K_s data-land use, and **e**, **f** K_s data-total

log-transformed K_s data, the AEP method did not reject the null hypothesis and they were approximately normally distributed. This was also observed from the skewness and kurtosis values, which are close to zero and the *Q-Q* plot shown in Fig. 1b, in which maximum data points lie on the straight line.

The CASF and PLAG land uses rejected the null hypothesis (data is normally distributed) at 5% significant level ($P < 0.05$) and hence, the K_s data (K_s data-land use) for these land uses were approximately not normally distributed. This is also evident from the skewness and kurtosis values, which are not close to zero and the normal *Q-Q* plot of PLAG given in Fig. 1c, for which most of the data points slightly away from the straight line. The K_s data for all other land uses including PADF and MANF did not reject the null hypothesis and distributed normally (Table 3). With the log-transformed K_s data, all land uses were not rejected the null hypothesis and they were approximately normally distributed. This is also evident from the skewness and kurtosis values, which are close to zero and the *Q-Q* plot of PLAG shown in Fig. 1d, in which maximum data points lie on the straight line.

The pooled K_s data (K_s data-total) rejected the null hypothesis (data is normally distributed) at 5% significant level ($P < 0.05$) and hence, the K_s data-total were approximately not normally distributed. This was also evident from the skewness and kurtosis values, which are not close to zero and normal the *Q-Q* plot given in Fig. 1e, in which the data points did not follow the trend of a straight line. With the log-transformed K_s data, K_s data-total did not reject the null hypothesis and they

Table 3 Normality test results for K_s data collected from four land uses using three K_s measuring techniques

Data group	Particulars (mm h^{-1})		Shapiro-Wilk			Skewness	Kurtosis
			Statistic	*df*	Sig.		
K_s methods	K_s	DRI	0.97	12	$	0.06	−0.39
		AEP	0.88		*	1.21	0.96
		PTF	0.93		$	0.05	−1.35
	$\ln K_s$	DRI	0.93		$	−0.75	−0.74
		AEP	0.96		$	0.5	−0.08
		PTF	0.91		$	−0.48	−1.1
K_s-landuses	K_s	PADF	0.94	9	$	0.13	−1.22
		MANF	0.95		$	0.48	−0.25
		CASF	0.79		***	1.6	1.8
		PLAG	0.87		*	1.27	1.41
	$\ln K_s$	PADF	0.91		$	−0.5	−1.07
		MANF	0.94		$	−0.3	−0.89
		CASF	0.9		$	0.98	0.37
		PLAG	0.95		$	0.43	−0.77
K_s-total	K_s		0.89	36	***	1.51	5.95
	$\ln K_s$		0.97		$	−0.25	−0.98

$ >0.05; *<0.05; **<0.01; ***<0.001

were approximately normally distributed (Table 3). This was also evident from the skewness and kurtosis values, which are close to zero and the *Q*-*Q* plot (Fig. 1f) in which maximum data points lie on the straight line.

It was understood that the K_s data were performed better according to the nature of log-normal frequency distribution as it was also observed by several other researchers [17, 20, 46]. Therefore, the log-transformed K_s data were used in studying the performance of the methods for different land uses on the basis of the descriptive statistics and ANOVA as explained in the previous section.

3.2 *Effect of Measuring Technique on* K_s

Table 4 presents the ANOVA of K_s data obtained from all land uses. The effect of the method on K_s was significantly different for all land uses at 5% significant level. Table 5 shows the pair-wise mean comparison of the measured/estimated K_s. Mostly, the mean K_s obtained by each method for a land use was significantly different from other methods. That means the methods measure different K_s values for a particular land use as different methods have different operating process and different sample dimensions. It showed that AEP method was highly variable with

Table 4 Two-way ANOVA test for the K_s data to analyze the effect of measuring technique and land uses on K_s

Source	Sum of squares (SS)	*df*	Mean square (MS)	*F*	Sig.
Method	12.49	3	3.12	45.92	***
Land use	6.53	3	1.63	24	***
Method × land use	2.80	8	0.14	2.05	**
Error	3.20	47	0.06		
Total	389.94	75			

*<0.05; **<0.01; ***<0.001

Table 5 Pair-wise comparisons the mean K_s of different methods

Method	Methods	Mean difference	Sig.	95% confidence interval	
(*i*)	(*j*)	(*i* − *j*)		Lower bound	Upper bound
DRI	AEP	−0.62	***	−0.79	−0.46
	PTF	0.39	***	0.23	0.56
AEP	DRI	0.62	***	0.46	0.79
	PTF	1.02	***	0.85	1.19
PTF	DRI	−0.39	***	−0.56	−0.23
	AEP	−1.02	***	−1.19	−0.85

*<0.05; **<0.01; ***<0.001

Table 6 Descriptive statistics of log-transformed K_s value for different methods

K_s-method	Statistics	Land use			
		PADF	MANF	CASF	PLAG
DRI	Mean (mm h^{-1})	2.17	1.79	2.46	1.51
	SD[a] (mm h^{-1})	0.04	0.3	0.13	0.49
	CV[b] (%)	0.02	0.17	0.05	0.32
	SE[c] (%)	7	0.05	0.04	0.07
	SV[d] (mm h^{-1})	0	0.09	0.02	0.24
	Range	0.08	0.54	0.26	0.94
AEP	Mean (mm h^{-1})	2.5	2.49	3.18	2.39
	SD[a] (mm h^{-1})	0.14	0.18	0.15	0.42
	CV[b] (%)	0.05	0.07	0.05	0.32
	SE[c] (%)	8	1	8	24
	SV[d] (mm h^{-1})	0.02	0.31	0.02	0.18
	Range	0.27	0.35	0.28	0.83
PTF	Mean (mm h^{-1})	0.85	1.63	2.13	1.36
	SD[a] (mm h^{-1})	0.22	0.27	0.15	0.32
	CV[b] (%)	0.25	0.16	0.07	0.24
	SE[c] (%)	12	16	9	19
	SV[d] (mm h^{-1})	0.05	0.08	0.02	0.11
	Range	0.39	0.51	0.28	20.62

[a]Standard deviation
[b]Coefficient of variation
[c]Standard error
[d]Sample variance

other methods as it rejected the null hypothesis at high level of significance. The PTF methods were not significantly different from each other which show the least difference between them.

The AEP method estimated K_s had high level of agreement with each other next to that of DRI method (Table 4). The PTF estimated K_s showed more variation and highly different from other methods. Furthermore, it can be interpreted that mean difference between the AEP and PTF methods were much significant. However the mean K_s was different for all methods mean K_s in DRI method (1.99 mm h^{-1}) is closely related to the overall mean of K_s for IIT Kharagpur campus (2.00 mm h^{-1}). The ranges of estimated/measured K_s were found to be higher for PTF method and comparatively lower in DRI method than the AEP methods (Table 6).

For all land uses, the mean K_s was maximum (2.64 mm h^{-1}) and minimum (1.59 mm h^{-1}) for AEP and PTF methods, respectively. The AEP method estimated highest values of K_s for all land uses which were ranged from 3.18 (CASF) to 2.39 mm h^{-1} (PLAG) with an average of 2.64 mm h^{-1} (SD = 0.22). Similarly, the DRI method resulted the second highest K_s values ranging from 2.46 (CASF) and 1.51 mm h^{-1} (PLAG) with an average of 1.99 mm h^{-1} (SD = 0.26). The Tukey's pair-wise comparison showed that there was no close correlation between the methods.

The descriptive statistics of estimated/measured K_s describing mean, standard deviation, coefficient of variation, standard error, sample variance and range were presented in Table 6. The estimated error was highest in PTF (28.20%) and lowest in case of DRI method (5.4%). The variance obtained for different methods implied that the highest was in AEP method (0.13 mm h^{-1}) and lowest in PTF method (0.06 mm h^{-1}). The higher K_s for AEP method was possibly due to the presence of higher water head and completely air-tight system [57, 58]. The PTF method of K_s measurement requires more accuracy in primary input data regarding the soil physical properties thus it may get erroneous result due to availability of inadequate soil data.

3.3 *Effect of Land Use on* K_s

The effect of the land use on K_s was significantly different for all methods (Table 4). Table 7 shows the pair-wise mean comparison of the measured/estimated K_s for the individual land uses. Mostly, the mean K_s of each land use considered was significantly different with that of other land use. That means the land uses had different values of K_s. However, the K_s was not significantly different for PADF, MANF and PLAG (Table 7).

The K_s varied between the land uses as they have different land management practices and presence of macrospores (created by earthworms, roots holes, etc.).

Table 7 Pair-wise comparisons of the mean K_s of different landuses

Land use	Land uses	Mean difference	Sig.	95% confidence interval	
(*i*)	(*j*)	(*i* − *j*)		Lower bound	Upper bound
PADF	MANF	−0.1	0.19	−0.25	0.05
	CASF	−0.66	***	−0.81	−0.5
	PLAG	0.09	0.24	−0.06	0.24
MANF	PADF	0.1	0.19	−0.05	0.25
	CASF	−0.56	***	−0.71	−0.4
	PLAG	0.19	*	0.03	0.34
CASF	PADF	0.66	***	0.5	0.81
	MANF	0.56	***	0.4	0.71
	PLAG	0.75	***	0.6	0.9
PLAG	PADF	−0.09	0.24	−0.24	0.06
	MANF	−0.19	*	−0.34	−0.03
	CASF	−0.75	***	−0.9	−0.6

$ >0.05; *<0.05; **< 0.01; ***<0.001

The K_s was lowest in PADY with 0.85 mm h^{-1} (PTF) and highest in CASF with 3.18 mm h^{-1} (AEP). The K_s for CASF was highest and was ranged from 2.13 (PTF) to 3.18 mm h^{-1} (AEP) with an average of 2.51 mm h^{-1} (SD = 0.15 mm h^{-1}). CASF had loose soil profile and large macropores (presence of earth worms) to result in higher values of K_s [59–61]. The PLAG showed the lowest K_s among all the land uses as it compacted regularly to make the field suitable for playing football and cricket [50, 62, 63]. From the textural analysis it was observed that clay percentage and bulk density were found higher in PADF, hence the measured K_s was found to be lower in PADF next to PLAG among all the land uses ranging from 0.85 (PTF) to 2.50 mm h^{-1} (AEP) with an average of 1.82 mm h^{-1} (SD = 0.10 mm h^{-1}). The bulk density of MANF was found to be higher next to PADF i.e. 1.78 g cm^{-3} and the measured K_s was also resulted closer to PADF. The K_s for MANF was ranging from 1.28 to 2.49 mm/h with an average value of 1.87 mm h^{-1} (SD = 0.19 mm h^{-1}). References [39, 64–66] were also found decrease in K_s with increase of bulk density and clay content which was similar as of our study for CASF and MANF land uses. The order of K_s obtained for the land uses were CASF (2.51 mm h^{-1}), MANF (1.87 mm h^{-1}), PADF (1.82 mm h^{-1}) and PLAG (1.71 mm h^{-1}) from higher to lower due to nature of porosity, organic and clay matter content, land management practices.

There was an obvious difference existed among the groups of K_s values for all four land uses by comparing pair-wise. The mean K_s of PADF (1.82 mm h^{-1}) and MANF (1.87 mm h^{-1}) were closely related. The mean K_s value of MANF (1.97 mm h^{-1}) is close match to the overall mean K_s (2.00 mm h^{-1}).

3.4 Feasibility of K_s *Methods*

Table 8 shows the description of the feasibility of a K_s measuring technique based on the equipment to measure, transportation-, installation- and operation times, water requirement, cost of instrument and operation, labour requirement, level of required skill and operation difficulties. Based on cost of operation, the maximum and minimum instrument cost was observed in AEP and PTF method, respectively. Operating time was observed to be lower in AEP methods among all the methods. Both the methods take equal time to reach steady state. The DRI methods were simple and easier and utilize same manpower and operating time. Proper installation with minimum soil disturbances for these instruments is important. The DRI method reduced the lateral flow of water from the inner ring which requires more water and resulted better measurement of K_s. The AEP method installation requires skilled labour and knowledge required to operate the instrument, and requires 30 min time for total operation of the instrument. The PTF method is an easier method among all which require less manpower, cost of operation but dependent upon accurate textural analyses. This is not surprise since each method has a characteristics different sampling volume, its own flow geometry, different method

Table 8 Feasibility of the K_s methods for different land uses

Performance criteria	DRI	AEP	PTF
Instrument/ equipment required	Two cylindrical rings, wooden piece leveller, hook gauge, stopwatch, scale, water	Cylindrical ring, reservoir, suction plate, stopwatch, scale, mug, water bucket, wooden piece leveller, manometer	None
Instrument cost ($)	250.00	280.00	None
Manpower	Skilled-1 Labour-1	Skilled-1 Labour-1	Skilled-1
Transportation	Laborious	Laborious	No
Installation time	30 min	45 min	No
Time to reach steady state	2–3 min	1–2 min	Not applicable
Operating time (h)	3–6 h	75 min	6–7 h
Water requirement (l)	40–50	25–35	2
Operation difficulties	Noisy installation and soil disturbance	Wetting front detection and noisy installation and soil disturbance	Textural analysis: chemical spill

of applying water to water to the soil and different boundary conditions. It can be realized that every method has its characteristics of sampling volume, own flow pattern or geometry, method of application of water to soil with certain initial and boundary conditions.

4 Conclusions

The variation of K_s was studied for different land uses of IIT Kharagpur campus using three different measurement techniques. The K_s data were approximately fitted well into log-normal distribution. The mean K_s was significantly different among the measuring or estimating techniques and the selected land uses. For all land uses, the mean K_s was highest for AEP followed by DRI and PTF methods. The estimated error in K_s was more for PTF and low for DRI methods. The PTF estimated K_s showed more variation and highly different from other methods. For the land uses, the mean K_s was highest for CASF followed by MANF, PADF and PLAG. In respect to land uses it was concluded that, the K_s measurement were highly affected by soil textural properties, porosity, clay content in the soil, bulk density and land management practices for PADF, MANF and PLAG land uses. Furthermore, the effect of interaction of method and land use on K_s was also significantly different at 5% level of significant. The selection of best suitable method for a particular situation can be obtained by optimizing the interdependent parameters, including method to be used, accuracy in instrument and measurement methods, soil condition and the numbers of practical constraints of the investigation (e.g. cost, availability of manpower, time requirement, portability of estimate, simplicity in measuring technique, operating condition). The overall performance of the methods suggested that the AEP method was quicker among all the methods.

References

1. Reynolds, W.D.: Saturated hydraulic conductivity: Laboratory measurement. In: Carter, M.R. (ed.) Soil Sampling and Methods of Analysis, pp. 589–598 (1993)
2. Shukla, M., Lal, R.: Transport of dissolve organic carbon through soil columns. Annual Meeting of ASA/SSSA Seattle, WA, p. 31 (2004)
3. Prieksat, M.A., Kaspar, T.C., Ankeny, M.D.: Positional and temporal changes in pounded infiltration in corn field. Soil Sci. Soc. Am. J. **58**, 181–184 (1994)
4. Lekamalage, W.B.: Characterization of surface soil hydraulic conductivity in sloping landscapes. Thesis Submitted to the College of Graduate Studies and Research in Partial Fulfillment of the Requirements for the Degree of Master of Science in the Department of Soil Science University of Saskatchewan Saskatoon, pp. 1–4 (2003)
5. Lee, H.J.: Comparing the inverse parameter estimation approach with pedo-transfer function method for estimating soil hydraulic conductivity. Geosci. J. **9**(3), 269–276 (2005)
6. Kechavarzi, C., Dawson, Q., Leeds-Harrison, P.B.: Physical properties of low-lying agricultural peat soils in England. Geoderma **154**, 196–202 (2010)
7. Jarvis, N., Koestel, J., Messing, I., Moeys, J., Lindahl, A.: influence of soil, land use and climatic factors on the hydraulic conductivity of soil. Hydrol. Earth Syst. Sci. **17**, 5185–5195 (2013)
8. Holden, J., Burt, T.P.: Hydraulic conductivity in upland blanket peat. Measur. Var. Hydrological Process. **17**, 1227–1237 (2003)
9. Rossiter, G.D., Jatten, G.V.: Effects of soil depth and saturated hydraulic conductivity spatial variation on runoff simulation by Limburg soil erosion model (LISEM). A Case Study in Faucon Catchment, France. Enschede, the Nederland's (2011)
10. Nielsen, D.R., Biggar, J.W., Erh, K.T.: Spatial variability of field-measured soil-water properties. Hilgardia **42**, 215–259 (1973)
11. Darzi, A., Yari, A., Bagheri, H., Sabe, G., Yari, R.: Study of variation of saturated hydraulic conductivity with time. J. Irrig. Drain. Eng. **134**, 479–484 (2008)
12. Bagarello, V., Provenzano, G., Sgroi, A.: Fitting particle size distribution models to data from burundian soils for the best procedure and other purposes. Biosys. Eng. **4**, 435–441 (2009)
13. Dev, K.S., Shukla, K.M.: Variability of hydraulic conductivity due to multiple factors. Am. J. Environ. Sci. **8**(5), 489–502 (2012)
14. Reynolds, D.W., Bowman, B.T., Brunke, R.R., Drury, C.F., Tan, C.S.: Comparison of tension infiltrometer, pressure infiltrometer, and soil core estimates of saturated hydraulic conductivity. Soil Sci. Soc. Am. J. **64**, 478–484 (2000)
15. Bagarello, B., Castellini, M., Di Prima, S., Giordano, G., Iovino, M.: Testing a simplified approach to determine field saturated soil hydraulic conductivity. Procedia Environ. Sci. **19**, 599–608 (2013)

16. Klute, A.: Laboratory measurement of hydraulic conductivity of saturated soil. In: Black, C.A. (ed.). Methods of Soil Analysis, Part 1, pp. 210–221 (1965)
17. Lee, M.D., Reynolds, D.W., Elrick, E.D., Clotheier, E.B.: A comparison of three field method for measuring saturated hydraulic conductivity. Can. J. Soil Sci. **65**, 563–573 (1985)
18. Klute, A., Dirksen, C.: Hydraulic conductivity and diffusivity: laboratory methods. In: Methods of Soil Analysis: Part 1—Physical and Mineralogical Methods, pp. 687–734 (1986)
19. Grant, C.D., Groenevelt, P.H.: Weighting the differential water capacity to account for declining hydraulic conductivity in a drying coarse-textured soil. Soil Res. **53**(4), 386–391 (2015)
20. Jačka, L., Pavlásek, J., Kuráž, V., Pech, P.: A comparison of three measuring methods for estimating the saturated hydraulic conductivity in the shallow subsurface layer of mountain podzols. Geoderma **219**, 82–88 (2014)
21. Burgy, R.H., Luthin, J.N.: A test of the single and double ring type infiltrometers. Trans. Am. Geophys. Union **37**, 189–191 (1956)
22. Bagarello, V., Iovino, M., Lai, J.B.: Field and numerical tests of the two-ponding depth procedure for analysis of single-ring pressure infiltrometer data. Pedosphere **23**(6), 779–789 (2013)
23. Elrick, D.E., Reynolds, W.D., Tan, K.A.: Hydraulic conductivity measurements in the unsaturated zone using improved well analyses. Groundw. Monit. Remediat. **9**(3), 184–193 (1989)
24. Bouwer, H.: A double tube method for measuring hydraulic conductivity of soil in situ above a water table. In: Soil Science Society of America Proceedings, vol. 25, pp. 334–342 (1961)
25. Perroux, K.M., White, I.: Designs for disc permeameters. Soil Sci. Soc. Am. J. **52**, 1205–1215 (1988)
26. Bouwer, H.: Rapid field measurement of air entry value and hydraulic conductivity of soil as significant parameters in flow system analysis. Water Resour. Res. **2**, 729–738 (1966)
27. Bagarello, V., Iovino, M., Elrick, D.: A simplified falling-head technique for rapid determination of field saturated hydraulic conductivity. Soil Sci. Soc. Am. J. **68**, 66–73 (2004)
28. Topp, G.C., Binns, M.N.: Field measurement of hydraulic conductivity with a modified air entry permeameter. Can. J. Soil Sci. **56**, 139–147 (1976)
29. Mohanty, B.P., Kanwar, S.R., Everts, J.C.: Comparison of saturated hydraulic conductivity measurement methods for a glacial-till soil. soil Sci. Soc. Am. J. **58**(3), 672–677 (1994)
30. Bodhinayake, W., Si, C.B.: Near-saturated surface soil hydraulic properties under different land uses in the St Denis National Wildlife Area, Saskatchewan, Canada. Hydrol. Process. **18**, 2835–2850 (2004)
31. Fallico, C., Migliari, E., Troisi, S.: Comparison of three measurement methods of saturated hydraulic conductivity. Hydrol. Earth Syst. Sci. Discuss. **3**, 987–1019 (2006)
32. Bagarello, V., Sgroi, A.: Using the single-ring infiltrometer method to detect temporal changes in surface soil field saturated hydraulic conductivity. Soil Tillage Res. **76**(1), 13–24 (2004)
33. Fodor, N., Sandor, R., Orfanus, T., Lichne, L., Rajkai, K.: Evaluation method dependency of measured saturated hydraulic conductivity. Geoderma **165**, 60–68 (2011)
34. Ronayne, M.J., Houghton, T.B., Stednick, J.D.: Field characterization of hydraulic conductivity in a heterogeneous alpine glacial till. J. Hydrol. Eng. **458**(459), 103–109 (2012)
35. Runbin, D.R., Fedler, B.C., Borrelli, J.: Comparison of methods to estimate saturated hydraulic conductivity in texas soils with grass. J. Irrig. Drain. Eng. **138**(4), 322–327 (2012)
36. Bagarello, V., Baiamonte, G., Castellini, M., Di Prima, D., Iovino, M.: A comparison between the single ring pressure infiltrometer and simplified falling head techniques. Hydrol. Process. **28**, 4843–4853 (2014)
37. Hall, D.G., Reeve, M.J., Thomasson, A.J., Wright, A.F.: Water Retention, Porosity and Density of Field Soils. Soil Survey of England and Wales. Rothamsted Experimental Station, Harpenden, UK (1977)
38. Campbell, G.S.: Soil Physics with Basic: Transport Models for Soil Plant Systems. Elsevier Science, New York (1985)

39. Rawls, W.J., Gimenez, D., Grossman, R.: Use of soil texture, bulk density and slope of the water retention curve to predict saturated hydraulic conductivity. Am. Soc. Agric. Biol. Eng. **41**(4), 983–988 (1998)
40. Smettem, K.R.J., Bristow, K.L.: Obtaining soil hydraulic properties for water balance and leaching models from survey data. 2. Hydraulic conductivity. Aust. J. Agric. Res. **50**(7), 1259–1262 (1999)
41. Wösten, J.H.M., Pachepsky, Y.A., Rawls, W.J.: Pedotransfer functions: bridging the gap between available basic soil data and missing soil hydraulic characteristics. J. Hydrol. **251**, 123–150 (2001)
42. Wosten, J.H.M., Lilly, A., Nemes, A., Le Bas, C.: Development and use of a database of hydraulic properties of European soils. Geoderma **90**, 169–185 (1999)
43. Schaap, M.G., Leij, F.J., Van Genuchten, M.T.: Rosetta: A computer program for estimating soil hydraulic parameters with hierarchical pedotransfer functions. J. Hydrol. **251**, 163–176 (2001)
44. Wagner, B., Tarnawski, V.R., Hennings, V., Müller, U., Wessolek, G., Plagge, R.: Evaluation of pedo-transfer functions for unsaturated soil hydraulic conductivity using an independent data set. Geoderma **102**, 275–297 (2001)
45. Bouma, J.: Measuring the hydraulic conductivity of soil horizons with continuous macropores. Soil Sci. Soc. Am. J. **46**, 438–441 (1983)
46. Nielsen, D.R., Wendroth, O.: Spatial and Temporal Statistics—Sampling Field Soils and Their Vegetation. Catena, Reiskirchen, Germany, p. 416 (2003)
47. Bormann, H., Klaassen, K.: Seasonal and land use dependent variability of soil hydraulic and soil hydrological properties of two northern german soils. Geoderma **145**, 295–302 (2008)
48. Hu, W., Shao, M., Wang, Q., She, D.: Effects of measurement method, scale, and landscape features on variability of saturated hydraulic conductivity. J. Hydrol. Eng. **18**, 378–386 (2013)
49. Reynolds, W.D., Elrick, D.E., Young, E.G.: Ring or cylinder infiltrometers (vadose zone). In: Dane, J.H., Topp, G.C. (eds.) Methods of Soil Analysis, Part 4: Physical Methods. Soil Science Society of America Journal Madison, pp. 818–843 (2002)
50. Hu, W., Shao, M., Wang, Q., Fan, J., Horton, R.: Temporal changes of soil hydraulic properties under different land uses. Geoderma **149**(3–4), 355–366 (2009)
51. Carter, M.R., Gregorich, E.G.: Soil Sampling and Methods of analysis. Canadian Society of Soil Science, Pinawa, Manitoba (2008)
52. Rao, M.D., Raghuwanshi, N.S., Singh, R.: Development of a physically based 1d-infiltration model for irrigated soils. Agric. Water Manag. **85**(1), 165–174 (2006)
53. Nimmo, J.R., Schmidt, K.M., Perkins, K.S., Stock, J.D.: Rapid measurement of field-saturated hydraulic conductivity for areal characterization. Vadose Zone J. **8**(1), 142–149 (2009)
54. Schaap, M.G.: Rosetta Version 1.0. U.S. Salinity Laboratory, ARS, U.S. Department of Agriculture, Riverside, CA. (1999)
55. Alvarez-Acosta, C., Lascano, R.J., Stroosnijder, L.: Test of the Rosetta pedotransfer function for saturated hydraulic conductivity. Open J. Soil Sci. **2**(3), 203–212 (2012)
56. Schaap, M.G., Van Genuchten, M.T.: A modified Mualem-Van Genuchten formulation. Vadose Zone J. **5**, 27–34 (2006)
57. Aldabagh, A.S.Y., Beer, C.E.: Field measurement of hydraulic conductivity above a water table with air-entry permeameter. Am. Soc. Agric. Biol. Eng. **14**(1), 29–31 (1971)
58. Nemati, M.R., Caron, J., Banton, O., Tardif, P.: Determining air entry value in peat substrates. Soil Sci. Soc. Am. J. **66**(2), 367–373 (2002)
59. Van Den Berg, J.A., Louters, T.: The variability of soil moisture diffusivity of loamy to silty soils on marl, determined by the hot air method. J. Hydrol. **97**(3), 235–250 (1988)
60. Chapuis, R.P.: Predicting the saturated hydraulic conductivity of sand and gravel using effective diameter and void ratio. Can. Geotech. J. **41**(5), 787–795 (2004)
61. Park, E., Smucker, A.: Saturated hydraulic conductivity and porosity within macro aggregates modified by tillage. Soil Sci. Soc. Am. J. **69**(1), 38–45 (2005)

62. Zhou, X., Lin, H.S., White, E.A.: Surface soil hydraulic properties in four soil series under different land uses and their temporal changes. CATENA **73**(2), 18–188 (2008)
63. Matthews, G.P., Laudone, G.M., Gregory, A.S., Bird, N.R.A., Matthews, A.G., Whalley, W. R.: Measurement and simulation of the effect of compaction on the pore structure and saturated hydraulic conductivity of grassland and arable soil. Water Resour. Res. 10.1029/2009WR007720 (2010)
64. Zhang, X.C., Norton, L.D.: Effect of exchangeable mg on saturated hydraulic conductivity disaggregation and clay dispersion of disturbed soils. J. Hydrol. **260**(1–4), 194–205 (2002)
65. Lado, M., Paz, A., Ben-Hur, M.: Organic matter and aggregate size interactions in saturated hydraulic conductivity. Soil Sci. Soc. Am. J. **68**(1), 234–242 (2004)
66. Dusa, A.A.: Effect of bulk density on saturated hydraulic conductivity. J. Eng. Appl. Sci. **5**(1), 159–165 (2013)

A Study on Assessment of Groundwater Quality at Certain Industrial Zones in Visakhapatnam, Andhra Pradesh

P. V. R. Sravya, T. P. Sreejani and G. V. R. Srinivasa Rao

Abstract The present work aims at the evaluation of groundwater quality at certain industrial areas in and around Bharat Heavy Electrical Limited (BHPV), Natural Thermal Power Corporation Ltd. (NTPC) and Hindustan Petroleum Corporation Ltd. (HPCL) in Visakhapatnam city, Andhra Pradesh. Ground water samples are collected from all the three industrial zones and subjected to physico-chemical analysis for various parameters such as pH, Turbidity, Conductivity, Total Acidity, Total Hardness, Chlorides using standards of APHA. Water Quality Index is calculated by using the popular NSFWQI method and the quality of the water is rated as unfit for drinking. The data is subjected to statistical analysis using SPSS 20.0 software. The statistical methods used for the data analysis are Cluster Analysis and Factor Analysis. Based on the WQI and Statistical analysis, the contamination is observed due to the presence of excess amounts of hardness and chlorides.

Keywords SPSS 20.0 · NSFWQI · CA · FA · Physico-chemical parameters

1 Introduction

Safe drinking water is a necessity for human beings to sustain their life. The use of water in agriculture, domestic purpose, industrial applications, construction purpose, etc., also necessitates good quality. Quality of water refers to the chemical, physical, biological, and radiological characteristics of water. The quality of groundwater gets affected when various impurities/pollutants join the groundwater table.

On-site sanitation systems, effluents from industrial activities, leaking sewers, excess amounts of fertilizers applied in agricultural activities and existence of toxic

P. V. R.Sravya
Civil Engineering Department, ANITS, Visakhapatnam, India

T. P. Sreejani · G. V. R. Srinivasa Rao (✉)
Civil Engineering Department, A U Engineering College, Andhra University, Visakhapatnam, India
e-mail: gvrsrao@gmail.com

M. Rathinasamy et al. (eds.), *Water Resources and Environmental Engineering I*,
https://doi.org/10.1007/978-981-13-2044-6_11

substances naturally are the major causes of contamination. The industrial effluents though small in quantities have a significant impact on the groundwater quality when they are improperly disposed off on to the land.

The water quality data can be summarized in terms of an index called WQI. In the mid-twentieth century, Horton [1] coined WQI for the first time. pH, specific conductance, alkalinity, chlorides, DO and Coliforms were considered to calculate WQI with the weightages ranging from 1 to 4. The summations of these indices were divided by the sum of weights and were multiplied by the coefficient of temperature and coefficient of obvious pollution. Prati et al. [2] used a parametric scale varying them from 0 to 13 and the parameters with values more than 8 were considered to denote heavy pollution. Later, Dinius [3] designed a social accounting system to measure the costs and impact of pollution control efforts and applied WQI to the data of various streams in Alabama, USA. Brown et al. [4] proposed multiplicative form of WQI where individual weights were assigned to the parameters. The assigned weights reflect the significance of the parameters impacting the index. Similar methodologies have been considered by various other researchers depending on the usage and parameters considered. In another method by Inhaber [5], different parameters were introduced to the two sub-indices, one dealing the industrial and domestic wastes and the other one dealing the background water quality. Equal weightages were given to both the sub-indices and were averaged to give the WQI ranging from zero to large numbers representing the best and worse qualities of water respectively. Water quality modeling mathematically simulates the prediction of water pollution, with formulations for the determination of the position and momentum of pollutants in a water body. Cluster Analysis is the most popular data analysis tool for giving solutions to classification problems. This sorts out cases into clusters, where the degree of association is strong between members of the same cluster and weak between members of different clusters (Panda et al. [6]; Ozbay et al. [7]). CA is useful for the analysis and the interpretation of surface water dataset for assessing the sources of pollution and to understand the temporal and spatial variations in the water quality parameters which in turn enables the optimization of the network of regional water quality management (Zare [8]; Fataei et al. [9]; Salah et al. [10]). The most common statistical approach that provides intuitive similarity relationships between any one sample and the entire data set is hierarchical agglomerative clustering which provides visual summaries of the clustering process. This is generally represented by a typical dendrogram [11, 12].

2 Study Area

Visakhapatnam is the largest industrial city of the Indian state of Andhra Pradesh. The city comprises of good number of industrial complexes of varying sizes, both in public and private sectors. Important industrial complexes include petrochemical complexes, steel plant, thermal power plant, bulk drugs and pharmaceutical complexes and other along with largest port of the state and dockyard. The industrial zone comprises of major industries like the Visakhapatnam Steel Plant, Bharat Heavy Plates & Vessels, Hindusthan Petroleum Corp. Ltd., Hindusthan Zinc Limited, Hindusthan Shipyard, Port Trust, L. G. Polymers, Coromandel Fertilizers and National Thermal Power Corporation. It also houses a number of medium and small scale industries. The adjoining areas of three major industrial complexes viz., BHPV, NTPC, and HPCL are taken into consideration as study area in this research work.

3 Methodology

The methodology is as follows (see Fig. 1).

4 Results and Discussions

The water samples collected from each industrial zone are analyzed for pH, Turbidity, Electrical Conductivity, Hardness, Acidity, Alkalinity, Chlorides. The results are used to derive the WQI to find out the water quality. The parametric values analyzed are shown in Table 1.

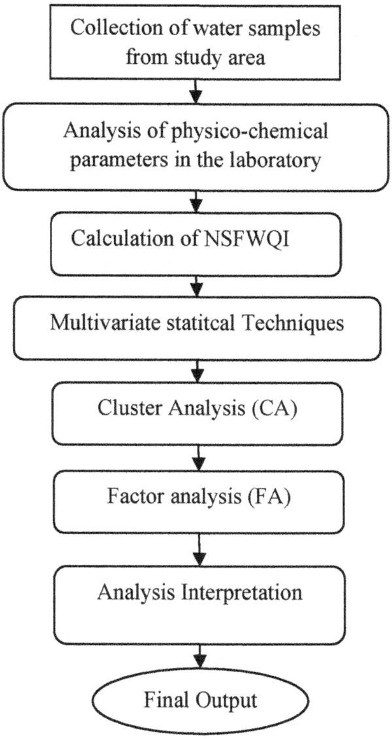

Fig. 1 Methodology adopted in the present study

Table 1 Results: water quality analysis

Industrial zone	Sample Nos.	Parameters						
		Total hardness (ppm)	Alkalinity (ppm)	Acidity (ppm)	Turbidity (ppm)	EC (μmho/cm)	pH	Chlorides (ppm)
BHPV	S1-1	434	510	18	8.9	1.48	8.12	174
	S1-2	468	372	4.5	8.4	1.22	9.27	276
	S1-3	890	766	36	6.5	2.94	7.9	450
	S1-4	346	494	27	4.5	1.61	8.13	404
	S1-5	403	673	34	0.8	1.71	8.1	141
	S1-6	307	467	32	10.2	1.93	7.78	159
	S1-7	489	439	31.5	9.4	1.8	7.95	301
	S1-8	569	429	22.5	6.4	2.17	7.8	226
	S1-9	432	423	45	5.3	1.89	8.01	209
	S1-10	571	363	31.5	9.7	1.8	7.77	170
NTPC	S2-1	210	461	9	2	1.27	8.5	436
	S2-2	520	515	13.5	9.3	1.11	8.45	368
	S2-3	751	483	31.5	8.1	2.41	8.18	251
	S2-4	440	314	45	7.1	3.85	8.26	762
	S2-5	444	439	22.5	9.8	1.5	7.7	244
	S2-6	394	393	18	1.57	4.7	8.42	365
	S2-7	934	314	40.5	6.9	1.5	8.3	1560
	S2-8	294	342	42.3	2	1.3	8.15	379
	S2-9	190	418	18	9.6	2.8	8.9	284
	S2-10	212	364	22	2.3	1.3	8.3	362
	S2-11	370	353	31.5	6.3	1	7.91	60
	S2-12	370	597	9	4.7	2.1	8.38	383
	S2-13	266	559	0	6.7	1.5	9.19	390
	S2-14	652	472	36	6.2	1.8	7.91	227
	S2-15	144	614	18	7.7	2.6	8.62	294
	S2-16	708	701	40.5	9.6	2.9	7.88	390
	S2-17	1052	657	54	7	3	7.8	454
	S2-18	190	324	37	6.4	3	8.06	451
HPCL	S3-1	261	428	18	9.6	0.14	8.06	195
	S3-2	499	504	36	5.2	1.63	8.03	304
	S3-3	302	233	27	4.4	1.04	7.73	219
	S3-4	616	210	9	8.8	0.63	8.57	393
	S3-5	403	448	31.5	7.1	1.57	7.88	134
	S3-6	491	477	27	6.8	1.12	8.17	78
	S3-7	413	428	22.5	8.5	1.52	8.04	358
	S3-8	285	225	45	6.3	0.93	7.73	81
	S3-9	338	442	27	5.7	1.65	7.85	127
	S3-10	509	477	22.5	5.7	1.62	8.72	230
	S3-11	531	396	40.5	5.6	1.38	8.06	304
	S3-12	613	494	26	2.3	1.6	8.69	485

4.1 Calculation of WQI

NSFWQI which ranges from 0 to 100 representing poor to excellent quality is calculated as given below.

$$\mathrm{WQI} = \sum \mathrm{SI}_i \tag{1}$$

$$\text{where } \mathrm{SI}_i = \mathrm{w}_i \times \mathrm{q}_i \tag{2}$$

SI_i sub-index of *i*th parameter
w_i relative weight and is calculated as

$w_i = 1/s_i$

s_i standard values of parameters
q_i quality rating of *i*th parameter
c_i experimental value.

The following table presents the WQI for all the three industrial zones (Table 2; Figs. 2, 3 and 4).

Table 2 WQI values at the industrial zones in Visakhapatnam

Sample No.	WQI at BHPV	WQI at NTPC	WQI at HPCL
1	140.97	69.65	143.30
2	74.20	121.36	215.17
3	221.45	205.20	165.9
4	168.06	267.81	96.74
5	187.87	166.03	199.76
6	214.42	112.15	176.93
7	209.42	246.68	161.52
8	152.65	233.17	262.18
9	259.54	144.87	171.78
10	209.99	134.33	151.55
11		196.2	
12		80.67	
13		45.45	
14		218.9	
15		137.2	
16		255.6	
17		312.1	
18		223.6	
Average	183.85	176.16	174.48

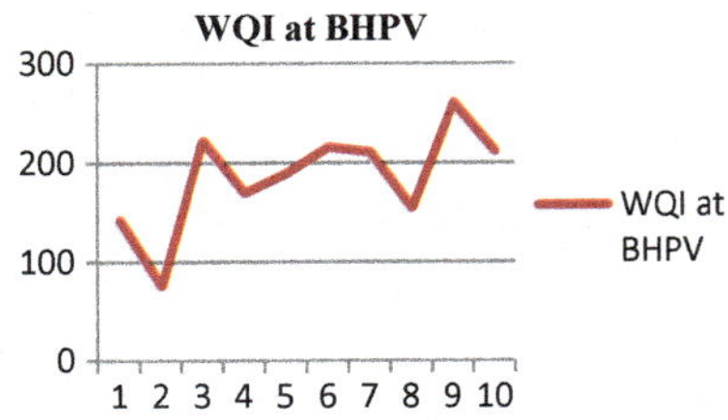

Fig. 2 Graphical representation of WQI at BHPV zone

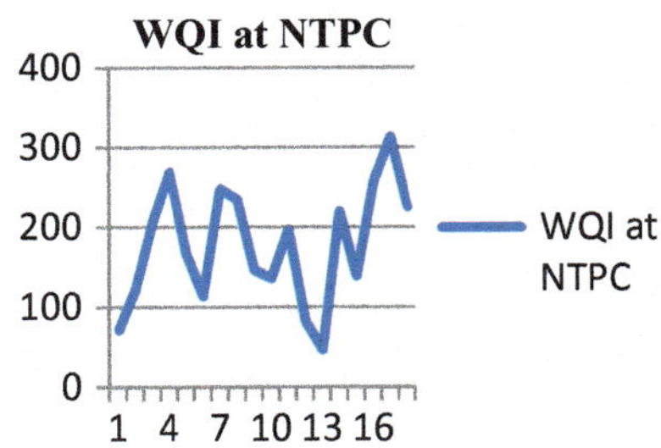

Fig. 3 Graphical representation of WQI at NTPC zone

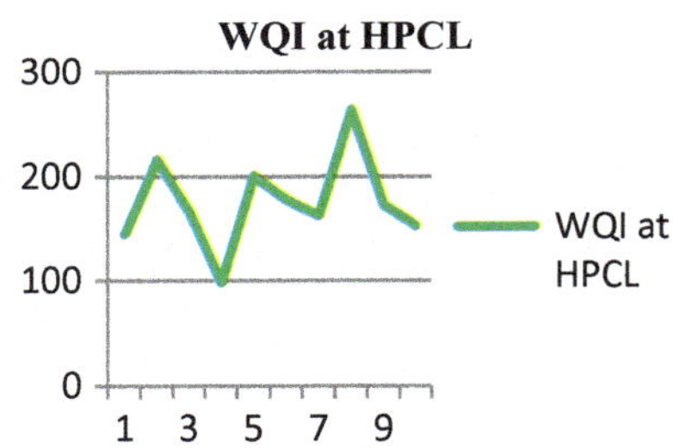

Fig. 4 Graphical representation of WQI at HPCL zone

4.2 Statistical Analysis

The CA of the water quality data of the study area resulted in the following agglomeration schedule as shown in Table 3. It gives the ida of the change in the distance measures of the clusters. The coefficients are the distance statistic forming the clusters. The sudden jump in the distance coefficient shows a good clustering phenomenon. A solution is the good solution that occurs before the gaps. For example, a large jump between cases 2 and 3 is observed in correspondence with the combination of cluster 1 (cases 4, 6), cluster 2 (4, 5) and cluster 3 (3, 4). The large jump between 4 and 5 forms a solution with two clusters i.e., cluster 1 (cases 1, 2), and cluster 2 (cases 1, 7). The agglomeration schedule shows that cases 4 and 6 form the cluster first at stage 1. Similarly, cases 4 and 5, 3 and 4, from clusters 2, 3 respectively. At stage 4, case 1 is combined with 2 to form cluster 2. At stage 5, cases 1 and 7 combined to form 2nd Cluster and cases 1 and 3 combined to form cluster 3 at stage 6.

Table 3 Agglomeration schedule for BHPV industrial zone

Stage	Cluster combined		Coefficients	Stage cluster first appears		Next stage
	Cluster 1	Cluster 2		Cluster 1	Cluster 2	
1	4	6	46.50	0	0	2
2	4	5	270.66	1	0	3
3	3	4	5001.00	0	2	6
4	1	2	113109.50	0	0	5
5	1	7	579324.33	4	0	6
6	1	3	3504953.71	5	3	0

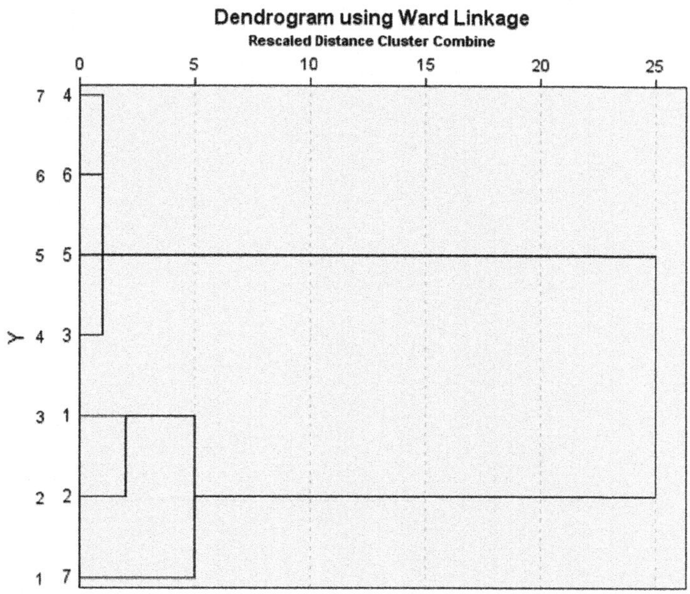

Fig. 5 Dendrogram for BHPV industrial zone

These rescaled distance coefficients and the cases are represented on *X* and *Y* axes respectively on the following Dendrogram see Fig. 5. It shows the closeness of the cases when they are combined. Herein, segment of the group is more for groups 2 and 7, i.e., the main pollutants are found to be Alkalinity and Chlorides.

The cluster solutions at the surroundings of NTPC industrial zone are presented in the following agglomeration schedule see Table 4. Based on the jumps in the coefficients the cluster solutions obtained are found to be in between the cases 1 and 2, 1 and 7, 3 and 4, 4 and 5 and 4 and 6. The segment group is more for the groups 2 and 7 and the main pollutants are alkalinity and chlorides (Fig. 6).

Table 4 Agglomeration schedule for NTPC industrial zone

Stage	Cluster combined		Coefficients	Stage cluster first appears		Next stage
	Cluster 1	Cluster 2		Cluster 1	Cluster 2	
1	4	6	91.50	0	0	2
2	4	5	420.66	1	0	3
3	3	4	9413.50	0	2	6
4	1	2	598184.00	0	0	5
5	1	7	1770216.16	4	0	6
6	1	3	8420179.42	5	3	0

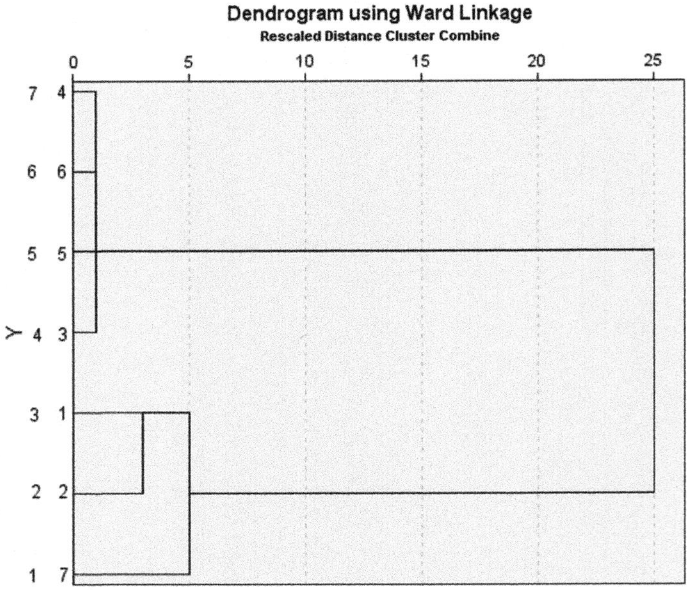

Fig. 6 Dendrogram for NTPC industrial zone

Similarly, the agglomeration schedule for the HPCL industrial zone see Table 5 shows the formation of the cluster solutions between the cases 1 and 2, 1 and 7, 3 and 4, 4 and 5, and 4 and 6.

The Dendrogram plot see Fig. 7 shows that the segment of the group is more for groups 2 and 7 and the main pollutants are alkalinity and chlorides.

The following tables depict the results of Descriptive statistics, Total Variance, Scree Plot (Figs. 8, 9 and 10). and Factor Matrix from the FA w.r.t. physico-chemical parameters at BHPV, NTPC, and HPCL industrial zones respectively.

Table 5 Agglomeration schedule for HPCL industrial zone

Stage	Cluster combined		Coefficients	Stage cluster first appears		Next stage
	Cluster 1	Cluster 2		Cluster 1	Cluster 2	
1	4	6	50.50	0	0	2
2	4	5	342.66	1	0	3
3	3	4	5649.25	0	2	6
4	1	2	129540.75	0	0	5
5	1	7	473920.58	4	0	6
6	1	3	3126559.71	5	3	0

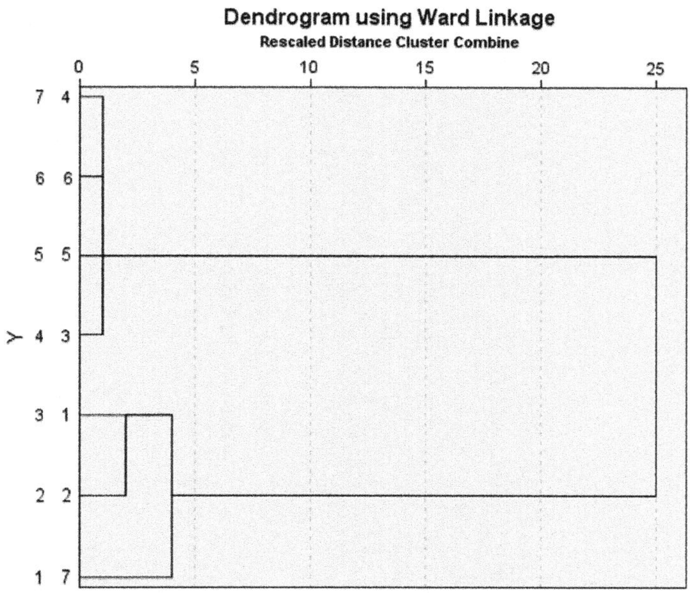

Fig. 7 Dendrogram for HPCL industrial zone

From the above-cited Table 6, it is observed that the highest mean and standard deviation values are obtained for the parameters Hardness, Alkalinity and Chlorides, that shows these have the greater influences on the quality of water.

The FA proved the following correlation matrices (see Tables 7, 8 and 9) for the different industrial zones.

The following Tables 10, 11 and 12 represent the percent of variance and the cumulative variance against different factors. The highest percentage variance is obtained from these tables from among the extracted values.

Table 6 Descriptive statistics (factor analysis at BHPV, NTPC and HPCL)

	Mean	Std. deviation	Analysis *N*	Mean	Std. deviation	Analysis *N*	Mean	Std. deviation	Analysis *N*
Hardness	**490.90**	**163.834**	10	**452.28**	**267.536**	18	**438.42**	**123.440**	12
Alkalinity	**493.60**	**129.677**	10	**462.22**	**122.522**	18	**396.83**	**109.402**	12
Acidity	28.20	11.302	10	27.11	14.724	18	27.58	9.774	12
Turbidity	6.80	2.936	10	6.33	2.787	18	6.33	2.146	12
EC	1.90	0.568	10	2.28	1.127	18	1.42	0.669	12
pH	8.10	0.316	10	8.17	0.383	18	8.25	0.452	12
Chlorides	**251.00**	**106.238**	10	**425.56**	**315.584**	18	**242.33**	**129.186**	12

The bold values represent the parameters which is exceeding the standard limits

Table 7 Correlation matrix (factor analysis at BHPV)

Parameters	Hardness	Alkalinity	Acidity	Turbidity	EC	pH	Chlorides
Hardness	1.000						
Alkalinity	0.462	1.000					
Acidity	0.122	0.330	1.000				
Turbidity	0.035	−0.529	−0.270	1.000			
EC	0.572	0.570	0.731	−0.280	1.000		
pH	−0.049	−0.329	−0.752	0.144	−0.557	1.000	
Chlorides	0.521	0.364	−0.010	−0.156	0.462	0.083	1.000

Table 8 Correlation matrix (factor analysis at NTPC)

Parameters	Hardness	Alkalinity	Acidity	Turbidity	EC	pH	Chlorides
Hardness	1.000						
Alkalinity	0.266	1.000					
Acidity	0.602	−0.170	1.000				
Turbidity	0.307	0.380	0.094	1.000			
EC	0.093	0.074	0.200	0.156	1.000		
pH	−0.434	0.256	−0.472	0.330	0.159	1.000	
Chlorides	0.412	−0.298	0.287	−0.005	0.126	−0.150	1.000

Table 9 Correlation matrix (factor analysis at HPCL)

Parameters	Hardness	Alkalinity	Acidity	Turbidity	EC	pH	Chlorides
Hardness	1.000						
Alkalinity	0.243	1.000					
Acidity	−0.240	0.046	1.000				
Turbidity	−0.239	−0.159	−0.443	1.000			
EC	0.355	0.473	0.126	−0.486	1.000		
pH	0.688	−0.017	−0.530	−0.187	0.225	1.000	
Chlorides	0.672	0.069	−0.0370	−0.251	0.243	0.593	1.000

From Table 13, it is observed that the hardness and chlorides have the higher the absolute values of the loadings contributing more to the variation in the quality of water.

Table 10 Total variance (FA at BHPV)

Factor	Initial eigen values			Extraction sums of squared loadings		
	Total	% of variance	Cumulative %	Total	% of variance	Cumulative %
1	3.16	45.23	45.23	3.02	43.226	43.226
2	1.61	23.11	68.34	1.38	19.723	62.949
3	1.07	15.37	83.72	0.82	11.791	74.740
4	0.51	7.30	91.03	0.27	3.920	78.660
5	0.34	4.88	95.91	0.10	1.427	80.087
6	0.19	2.75	98.66			
7	0.09	1.33	100.00			

Extraction method: principal axis factoring

Table 11 Total variance (FA at NTPC)

Factor	Initial eigenvalues			Extraction sums of squared loadings		
	Total	% of variance	Cumulative %	Total	% of variance	Cumulative %
1	2.25	32.23	32.23	1.75	25.12	25.12
2	1.75	25.11	57.34			
3	1.10	15.78	73.12			
4	0.81	11.70	84.82			
5	0.60	8.60	93.43			
6	0.32	4.56	98.00			
7	0.14	1.99	100.00			

Table 12 Total variance (FA at HPCL)

Factor	Initial eigenvalues			Extraction sums of squared loadings			Rotation sums of squared loadings		
	Total	% of variance	Cumulative %	Total	% of variance	Cumulative %	Total	% of variance	Cumulative %
1	2.799	39.992	39.992	2.644	37.776	37.776	1.680	23.998	23.998
2	1.861	26.585	66.578	1.636	23.376	61.152	1.038	14.833	38.831
3	0.996	14.222	80.799	0.652	9.314	70.466	1.036	14.807	53.638
4	0.492	7.028	87.827	0.319	4.557	75.023	1.026	14.659	68.297
5	0.383	5.477	93.304	0.224	3.197	78.219	0.695	9.923	78.219
6	0.343	4.895	98.199						
7	0.126	1.801	100.000						

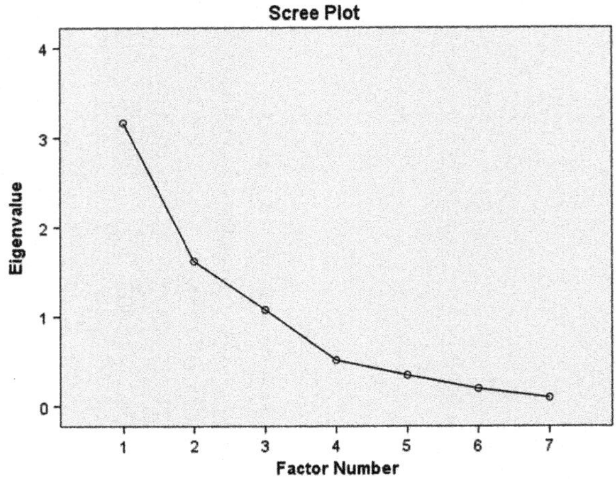

Fig. 8 Scree plot (FA at BHPV)

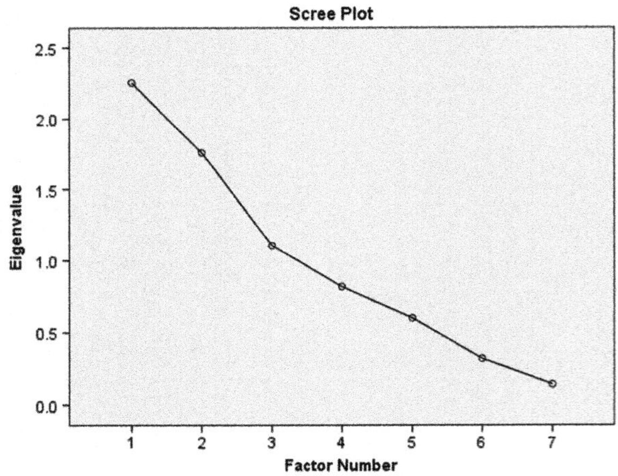

Fig. 9 Scree plot (FA at NTPC)

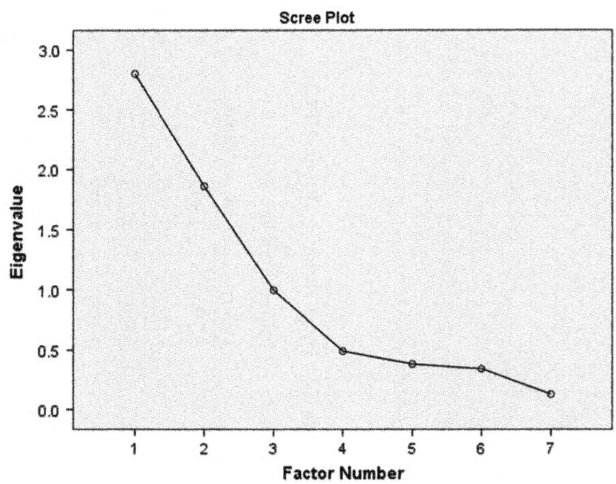

Fig. 10 Scree plot (FA at HPCL)

Table 13 Factor matrix at BHPV, NTPC and HPCL industrial zones

Parameters	BHPV					NTPC	HPCL				
	1	2	3	4	5	1	1	2	3	4	5
Hardness	**0.541**	0.806	0.332	0.153	0.146	**0.806**	**0.875**	0.048	0.011	0.361	0.158
Alkalinity	0.742	−0.080	−0.388	−0.306	−0.018	−0.080	0.237	0.384	0.585	0.019	0.013
Acidity	0.752	0.797	0.136	0.121	0.147	0.797	−0.424	0.780	−0.268	0.208	0.130
Turbidity	−0.425	0.080	0.693	−0.153	−0.056	0.080	−0.336	−0.709	0.323	0.194	0.100
EC	0.943	0.123	0.211	0.194	−0.036	0.123	0.449	0.548	0.273	−0.180	−0.011
pH	−0.619	−0.511	−0.141	0.187	0.166	−0.511	0.854	−0.261	−0.200	−0.233	0.246
Chlorides	**0.405**	0.430	0.034	0.214	0.157	**0.430**	0.773	−0.083	−0.136	0.146	−0.333

The bold values represent the parameters which is exceeding the standard limits

5 Conclusions

1. The average WQI for BHPV, NTPC and HPCL industrial zones are found to be 83.86, 176.17, and 178.28 respectively which shows that the water is not fit for drinking purposes directly.
2. Based on factor analysis, it can be concluded that hardness and chlorides have the greater influences on the quality of water with 45.23, 32.23, and 39.99% at all the three industrial zones viz., BHPV, NTPC and HPCL.
3. Cluster Analysis (CA) conducted at all the three zones has shown that Chlorides and Alkalinity are the major parameters affecting the degree of pollution.

References

1. Horton, R.K.: An index number system for rating water quality. J. Water Pollut. Control Fed. **37**(3), 300–306 (1965)
2. Prati, L., Pavanello, R., Pesarin, F.: Assessment of surface water quality by a single index of pollution. J. Water Res. **5**, 74 (1971)
3. Dinius, S.H.: Design of an index of water quality. Water Res. Bull. **23**(5), 833–843 (1987)
4. Brown, R.M., McLelland, N.I., Deininger, R.A., O'Connor, M.F.: A water quality index—crashing the psychological barrier. In: Indicators of Environmental Quality (1972)
5. Inhaber, H.: An approach to a water quality index for Canada. J. Water Res. **5**, 821 (1975)
6. Panda, U.C., Sundaray, S.K., Rath, P., Nayak, B.B., Bhatta, D.: Application of factor and cluster analysis for characterization of river and estuarine water systems—a case study: Mahanadi river (India). Elsevier—J. Hydrol. **331**, 434–445 (2006)
7. Ozbay, N., Yerel, S., Ankara, H.: Investigation of cluster analysis in surface water in Yesilirmak river. In: 1st International Symposium on Sustainable Development, Sarajevo, 9–10 June 2009
8. Zare Garizi, A.: Assessment of seasonal variations of chemical characteristics in surface water using multivariate statistical methods. Int. J. Environ. Sci. Tech. **8**(3), 581–592 (2011)
9. Fateai, E., Mosavi, S., Imani, A.A.: Identification of anthropogenic influences on water quality of Aras river by multivariate statistical techniques. In: 2nd International Conference on Biotechnology and Environment Management IPCBEE, vol. 42, pp. 35–39 (2012)
10. Salah, E.A.M., Turki, A.M., Al-Othman, E.M.: Assessment of water quality of Euphrates river using cluster analysis. J. Environ. Prot. **2**, 1629–1633 (2012)
11. Tokatli, C., Kose, E., Cicek, A., Emiroglu, O., Bastatli, Y.: Use of cluster analysis to evaluate surface water quality: an application from downstream of Meric river basin (Edirne, Turkey). Int. J. Adv. Sci. Eng. Technol. (Spl. Issue-3), 33–35 (2015) (ISSN: 2321-9009)
12. Ledesma-Ruiz, R., Pastén-Zapata, E., Parra, R., Harter, T., Mahlknecht, J.: Investigation of the geochemical evolution of groundwater under agricultural land: a case study in northeastern Mexico. Elsevier—J. Hydrol. **521**, 410–423 (2015)

1D and 2D Electrical Resistivity Investigations to Identify Potential Groundwater Resources in the Hard Rock Aquifers

L. Surinaidu

Abstract Hard rock aquifers develop significant groundwater potentials under favorable secondary porosity such as in fracture and fissures. However, identification of those zones is difficult due to their hydrogeological heterogeneities. The electrical resistivity methods are promising tools to identify the potential groundwater zones. In the present study, 1D and 2D electrical resistivity methods in the form of vertical electrical resistivity (VES) and electrical resistivity tomography (ERT) have been deployed for identifying potential groundwater zones in the Koppal district, Karnataka. The comparison of 1D and 2D resistivity results indicated that top soil, weathering zone, fissured and fractured zone and hard rock possessing resistivity values respectively <23, <230, <400 and >2000 ohm m. The low resistivity value <490 ohm m below hard rock in the area indicates the possibility of fracture zone underlying the hard rock. The weathered and fissured zone of 30 m thick is identified by a resistivity range of ~80 ohm m indicates a good groundwater potential in the study area. The location having weathering thickness of 10–20 m has been recommended for bore well drilling in the study which may yield about 200 m^3/day.

Keywords Electrical resistivity tomography · Vertical electrical sounding
Groundwater · Karnataka · Hard rocks · Aquifer

1 Introduction

The occurrence and distribution of groundwater resources in any region is predominantly controlled by rainfall pattern, geomorphology and geology in the area. The yield and occurrence of groundwater in the hard rock aquifers is entirely different than unconsolidated/alluvial formations due to its highly heterogeneous nature in the subsurface. The potential groundwater may develop in secondary

L. Surinaidu (✉)
CSIR-National Geophysical Research Institute (NGRI), Hyderabad, Telangana, India
e-mail: L.Surinaidu@ngri.res.in

M. Rathinasamy et al. (eds.), *Water Resources and Environmental Engineering I*,
https://doi.org/10.1007/978-981-13-2044-6_12

porosity such as fractures and fissures in the hard rock regions. It is evident that good amount of water may reside preferentially in the considerable thickness of weathering or fracture/jointing aquifer materials. Very often good quality of groundwater with high yielding wells can be found in the places of fractures and joints. The identification of such sources are very difficult. Electrical resistivity methods are promising techniques for different applications viz., identifying potential groundwater zones, delineation of groundwater contaminations, seawater intrusion mapping and other geotechnical applications [1–3]. Among all geophysical techniques 1D vertical electrical resistivity (VES) and 2D electrical resistivity tomography (ERT) are more popular to identify the potential groundwater zones, to demarcate aquifer geometry and to delineate the contaminate aquifer zones. These techniques basically use the subsurface resistivity anomalies that is driven by porosity, moisture/saturation levels and quality of moisture held in the formation to understand the quantity and quality of groundwater in the aquifer. Many researchers has applied electrical resistivity techniques for indentifying the potential aquifers [4], to delineate contaminate zones [5], to demarcated the saline water zones [6, 7] and for the locating the basement rocks [8].

The objective of the present study is to understand the aquifer geometry and to identify the potential groundwater zones using Electrical Resistivity Tomography (ERT) and Vertical Resistivity Sounding (VES) techniques.

2 Study Area

The present study is carried out for Hare Krishna Metallic Pvt. Ltd. (HKMPL) located in Hirebaganal village, Koppal district, Karnataka. The water sourcing for the factory operation has been made through two tube wells within their premises. HKMPL has plans to expand the operations by starting a co-generation thermal power plant, which would require additional water for cooling and other operations (Fig. 1). In the study area, a total of 14 vertical electrical soundings (VES) and 6 electrical resistivity tomography (ERT) investigations were carried out to find out the potential groundwater pockets. The average annual rainfall in the area is about 1100 mm and 80% of the rainfall received during the NE monsoon. Tungabhadra reservoir forms the southern part of the study area. Geomorphologically the area landforms are characterized by structural plateaus, hills and valleys, denudational plateaus, pediment/Pedi plain. The general groundwater levels in the area range from 18 to 21 m bgl (below ground level). The drainage pattern is dendritic to parallel in nature in the area. In general, the groundwater has been restricted to a depth of 60 m in weathered granitic rocks. The double ring infiltration tests in the area revealed that infiltrations rates are vary from 0.25 to 1.69 mm/h.

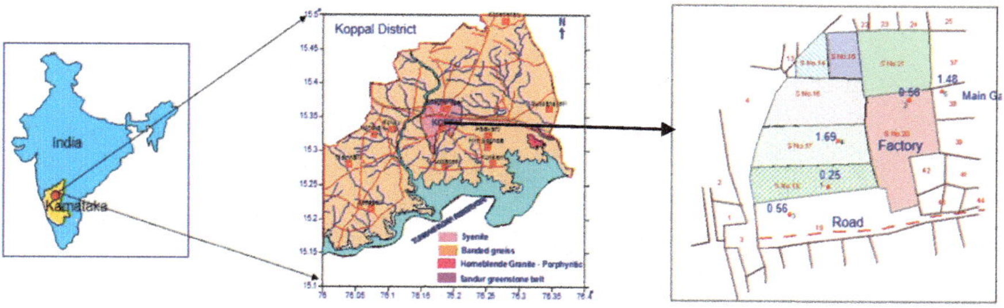

Fig. 1 Location map of the study area (*Note* S. No indicates survey number and number in blue color is infiltration rate mm/h)

3 Geology

The area forms pre-Proterozoic denudation plateaus, denudational hills and banded gneisses that belong to Archean age with hard and massive nature (Fig. 1). Detailed hydrogeological survey in the region indicates that the region is encompasses the rocks of peninsular Gneissic complex, Dharwar Super Group, Kaladgi super group and other minor intrusive rocks. Rocks belonging to Peninsular Gneissic Complex are exposed in Koppal taluka that include several bands and isolated patches of medium grained banded biotite gneiss. Younger granite is represented by gray biotite, hornblende granite, pink granite, and syenite. Grey biotite granite is massive in nature, occasionally porphyritic and contains patches of migmatitic gneiss. In the area, prominent outcrop of syenite is also seen around Koppal. Dykes of gabbro and dolerite are representing the last phase of intrusive igneous activity criss-cross the rocks in the area.

4 Methodology

4.1 Vertical Electrical Sounding (VES)

VES using Schlumberger electrode array have been deployed at 17 locations (Fig. 2). The half current electrode separation (AB/2) was increased incrementally starting from 1.5 to 150 m. The automated forward and inverse modeling software programs viz., WINSEV and RESIST for data processing and interpretation [9]. The apparent resistivity values versus electrode spacing (AB/2) obtained from VES entered as an input the In order to reduce the number of layers in VES generated by program an edge preserving and smoothing techniques is applied. The inverse modeling is initiated with estimated resistivities and thickness of layers. The program produces best fits model in the least square sense with ridge regression by iteratively adjusting the parameters of the starting model.

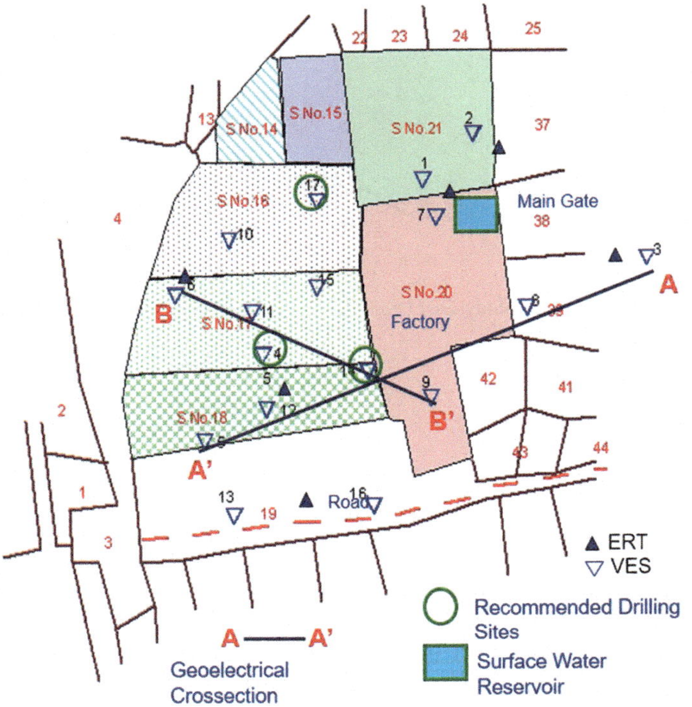

Fig. 2 Location of VES and ERT in the study area

4.2 *Electrical Resistivity Tomography*

The ERT surveys were carried out at six locations in and around the Hare Krishna Metallics, Koppal district, Karnataka (Fig. 2). The apparent resistivity data was collected using SYSCAL Pro resistivity meter with 48 electrodes using 5 m electrode spacing with Wenner-Schlumberger, Wenner and dipole–dipole configurations to delineate aquifer geometry and to identify the potential groundwater zones. RES2D.INV inversion software was used to convert the apparent resistivity values onto true resistivity [10]. Apparent resistivity values are calculated using finite-difference method [11] method.

The data inversion was carried out with the least squares inversion method [12].

5 Results and Discussion

5.1 VES

The interpreted geoelectric layers from 17 VES in the study area revealed that three lithological units that are top soil, weathered material and basement rock. The thickness of top soil varying from 0.5 to 1.2 m from the surface followed by thick weathered formation ranging from 10 to 33 m in the subsurface from the surface followed by hard basement. The interpreted geoelectric cross sections connecting KS06, KS11, KS04, KS 14 and KS09 along AA^1 and connecting KS03, KS06, KS14, KS12, and KS05 along BB^1 are shown in Fig. 3. The two geoelectric cross sections are shown that three distinctive subsurface layers in the area.

5.2 Correlation of ERT and VES

The ERT Nos. 1 and 2 have been carried out in the northern part of the Factory in Survey No. 21. Comparison of inverse vertical Resistivity sections computed from the ERT1 No. 1 and VES No. 2 results indicated a resistivity of <230 ohm m with 10 m thick is fissured and joint zone, a characteristic of weathering zone in the area. It is followed by 15 m thickness of fractured zone with Resistivity up to 580 ohm m underlain by in the granitic terrain at shallow depth. The hard rock in the region can be characterized with resistivities >1800 ohm m (Fig. 4).

Comparison of computed inverse model Resistivity Section of ERT No. 2 with VES No. 1 indicate top soil with 3 m thickness followed by 3.2 m of fissured and fractured zone of Resistivity ~400 ohm m. About 21 m of hard rock has been

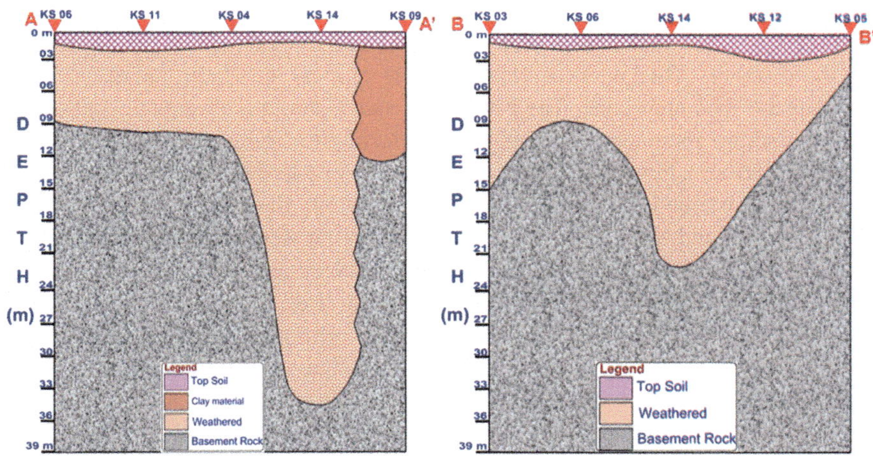

Fig. 3 Geoelectrical cross section along the profile AA^1 and BB^1 (refer Fig. 2 for location)

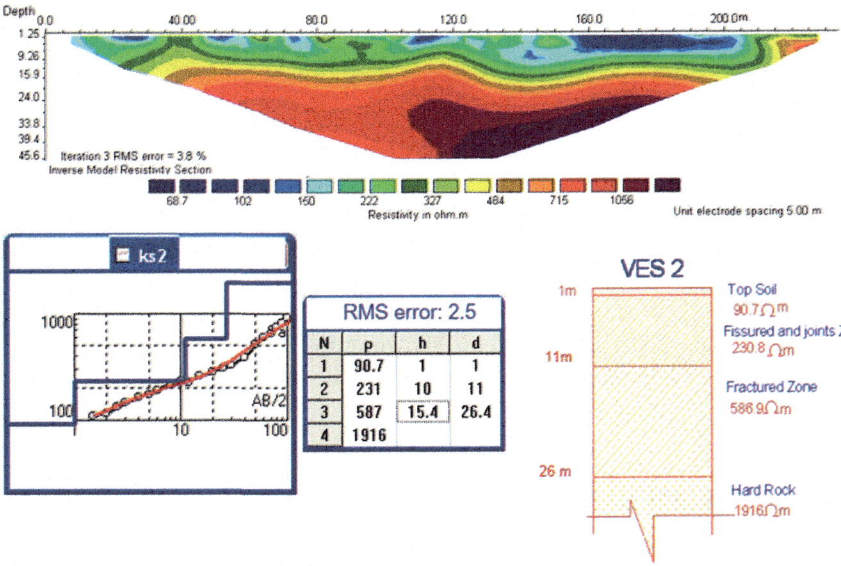

Fig. 4 Correlation of multi-electrode resistivity image no. 1 with VES no. 2 in the north side of the Hare Krishna metallics, Koppal district

encountered with Resistivity >2000 ohm m and at further depth the Resistivity has shown a low value 490 ohm m indicating possibility of fracture zone underlying the hard rock at this location (Fig. 5).

ERT No. 3 and VES No. 3 have been carried out in front of the main gate. The weathered zone posses a Resistivity ~50 ohm m and extends to 3.1 m depth followed by fractured zone with a very low Resistivity values <200 ohm m extending up to 18 m indicating a good groundwater potential zone outside the factory (Fig. 6).

ERT No. 4 and VES 6 have been carried out in Survey No. 27 in the western part of the Factory. Thickness of fracture zone is >30 m which indicate a good groundwater potential zone. The Resistivity values are in the range of 590 ohm m for the fracture zone and about 100 ohm m for the overlying weathered zone. The hard rock possess Resistivity values >1300 ohm m. Geoelectrical layers also have been found stratified with uniform variation all along the profile (Fig. 7).

ERT No. 5 and VES No. 4 was carried out on the western part of the factory in Survey No. 17. The combined thickness of weathered and fractured zone is hardly 10 m only with Resistivity value >590 ohm m (Fig. 8). ERT No. 6 and VES No. 14 have been carried out at outside the factory along the western boundary of the factory. The comparison of Resistivity values indicated weathered and fissured zone having a resistivity of ~80 ohm m, which is indicates a good groundwater potential area. The combined thickness of weathered and fissured zone in this case is >30 m, which can store and transmit good amount groundwater (Fig. 9). The site has been recommended for drilling of tube well, which may yield ~200 m^3/day.

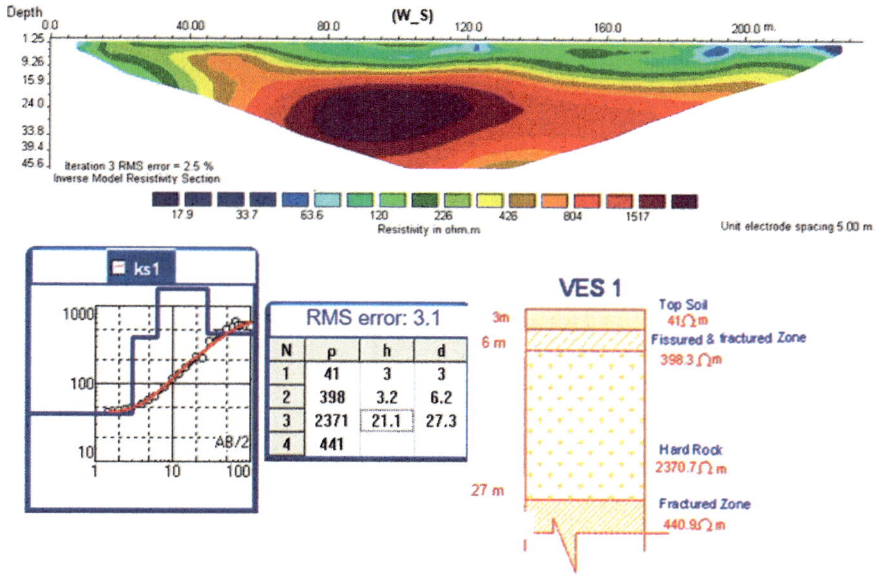

Fig. 5 Correlation of multi-electrode resistivity image no.2 with VES no. 1 in the north side of the Hare Krishna metallics, Koppal district

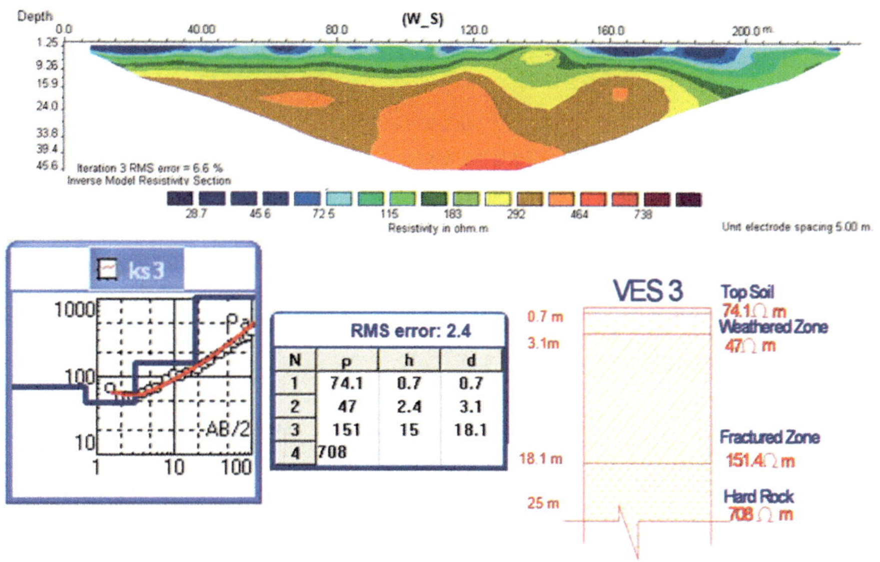

Fig. 6 Correlation of multi-electrode resistivity image no. 3 with VES no. 3 in the west side of the Hare Krishna metallics, Koppal district

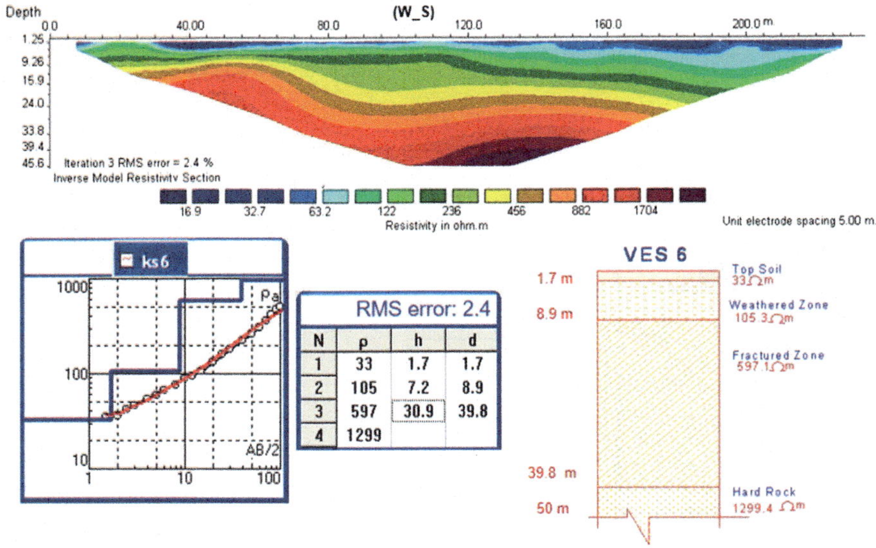

Fig. 7 Correlation of multi-electrode resistivity image no. 4 with VES no. 6 in the west side of the Hare Krishna metallics, Koppal district

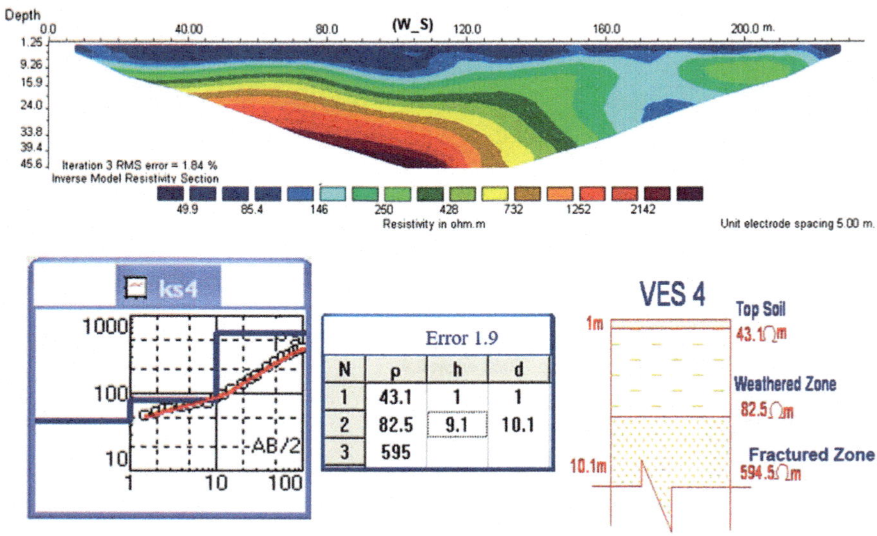

Fig. 8 Correlation of multi-electrode resistivity image no. 5 with VES no. 4 in the south west side of the Hare Krishna metallics, Koppal district

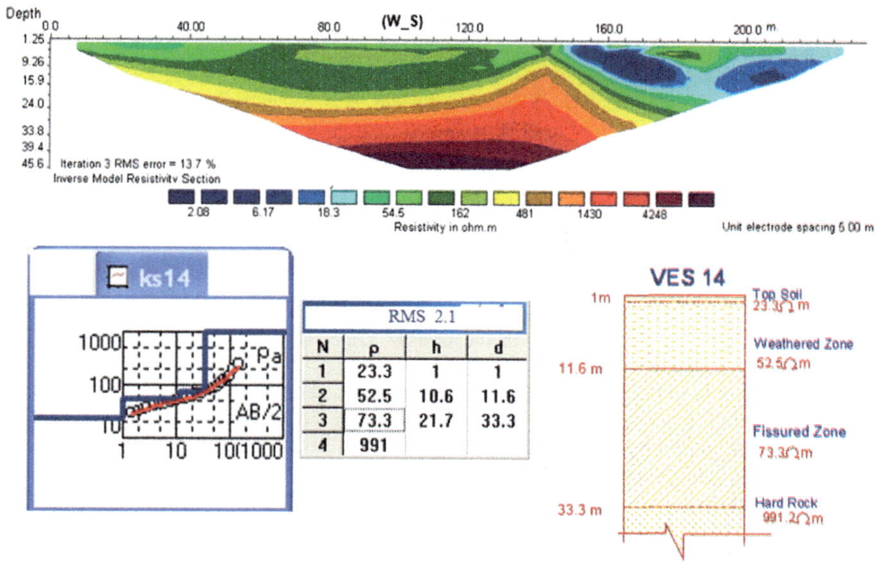

Fig. 9 Correlation of multi-electrode resistivity image no. 6 with VES no. 14 in the south side of the Hare Krishna metallics, Koppal district

6 Conclusion

Electrical resistivity techniques in the form of VES and ERT are successfully applied in the study area to identify the potential groundwater zones. The comparison of six inverse model Resistivity sections and inferred geoelectrical layers from 16 VES indicate that a very good groundwater potential exists in the western boundary of the study area and has been suggested for drilling up to 100 m depth. Rest of the locations has low to moderate groundwater potential is expected.

Acknowledgements I express my sincere thanks to the Director, CSIR-National Geophysical Research Institute (NGRI), Hyderabad for his kind permission to publish this paper. I am grateful to Dr. V. V. S. Gurunadha Rao for his encouragement to publish as a single author. I also would like to thank Mr. M. Durga Prasad for his help in preparing Fig. 3.

References

1. Griffiths, D.H., Barker, R.D.: Two-dimensional resistivity, imaging and modelling in areas of complex geology. Appl. Geophys. **29**, 211–226 (1993)
2. Telford, W.M., Geldart, L.P., Sheriff, R.E., Keys, D.A.: Applied Geophysics, 2nd edn., p. 700. Cambridge University Press, Cambridge (1976)
3. Loke, M.H.: Electrical Imaging Surveys for Environmental and Engineering Studies. Unpublished report, pp. 1–8 (1997)

4. Asry, Z., Samsudin, A.R., Yaacob, W.Z., Yaakub, J.: Groundwater investigation using electrical resistivity imaging technique at Sg. Udang, Melaka, Malaysia. Bull. Geol. Soc. Malays. **58**, 55–58 (2012)
5. TammaRao, G., Gurunadha, V.V.S., Surinaidu, L., Ranganadhan, K.: Assessment of groundwater contamination from a hazardous dumpsite in Ranipet, Tamil Nadu, India. Hydrogeol. J. **19**(8), 1587–1598 (2011). https://doi.org/10.1007/s10040-011-0771-9
6. Surinaidu, L., Rao, V.V.S.G., Rao, G.T., Mahesh, J., Padalu, G., Sarma, V.S.: An integrated approach to investigate saline water intrusion and to identify the salinity sources in the central Godavari delta, Andhra Pradesh, India. Arab. J. Geosci. **6**(10), 3709–3724 (2012). https://doi.org/10.1007/s12517-012-0634-2
7. Surinaidu, L., Rao, V.V.S.G., Prasad, P.R., Sarma, V.S.: Use of geophysical and hydrochemical tools to investigate seawater intrusion in coastal alluvial aquifer, Andhra Pradesh, India. In: Groundwater in the Coastal Zones of Asia-Pacific, vol. 7, pp. 49–65. Springer (2013)
8. Badmus, B.S., Olatinsu, O.B.: Geophysical characterization of basement rocks and groundwater potentials using electrical sounding data from Odeda quarry site, south-western, Nigeria. Asian J. Earth Sci. **5**, 79–87 (2012)
9. Koefoed, O.: Geosounding Principle, vol. 1, p. 276. Elsevier, Amsterdam (1979)
10. Loke, M.H., Barker, R.D.: Practical techniques for 3D resistivity surveys and data inversion. Geophys. Prospect. **44**, 499–523 (1996b)
11. Dahlin, T.: 2D resistivity surveying for environmental and engineering applications. First Break **14**, 275–284 (1996)
12. Dey, A., Morrison, H.F.: Resistivity modelling for arbitrary shaped two-dimensional structures. Geophys. Prospect. **27**, 1020–1036 (1979)

Evaluation of Utilization of Wavelet Denoising Approach in Calibration of Hydrological Models

Maheswaran Rathinasamy, Akash Choudary and Anuj Jaiswal

Abstract Hydrological modeling can be very useful in studying the hydrology of the system and managing the water resources of the system in a sustainable way. Calibration of the hydrological model is an important step in model development and application. Calibration becomes difficult particularly when the input variables of the model is of poor quality and contaminated with noise. In order to improve the calibration and aid in modeling, denoising of the data has been used in past. In this study, a hydrological model for the Wainganga basin, India using SWAT coupled with wavelet denoising is developed. The model performance of the wavelet coupled SWAT model is compared with the simple SWAT model. For the purpose of the model calibration 8 years were used and the model validation was done using 3 years of data. The results from the study show that the wavelet-based denoising significantly improved the model performance and also aided model calibration.

Keywords Wavelet denoising · SWAT model · Model calibration

1 Introduction

Hydrological models are being extensively used to study large catchments and water resources management problems. Hydrological models are conceptual representations of a part of the hydrological cycle. They use inputs which are properties related to the actual catchment characteristics and simulate a particular process of the hydrological cycle such as groundwater, surface runoff, contaminant transport, etc.

M. Rathinasamy (✉)
Department of Civil Engineering, MVGR College of Engineering, Vizianagaram, Andhra Pradesh, India
e-mail: maheswaran27@yahoo.co.in

A. Choudary · A. Jaiswal
Department of Civil Engineering, IIT Delhi, New Delhi, India

M. Rathinasamy et al. (eds.), *Water Resources and Environmental Engineering I*,
https://doi.org/10.1007/978-981-13-2044-6_13

One of the most widely used models to simulate surface runoff is SWAT (Soil and Water Assessment Tool) [1–3]. This model uses inputs like ground elevation, soil types, land use, precipitation and temperature to simulate the surface runoff of a given basin. By default the model computes the values of certain parameters from the inputs and the model also need some parameters which cannot be related to the physical characteristics of the catchment. Calibration is usually done manually in which the user tries different sets of values in order to get the desired objective function (usually to reduce the error between the observed and simulated). The calibration and performance of the models depends on many underlying factors—accuracy of the input series, model conceptualization, etc. Many times, even though equipped with proper model conceptualization, still the model outputs may not be matching with the observed due to the error or noise in the input variables. Also, true measurements of these hydro-meteorological variables across full time and space domain are rarely available.

Even though there is a lot of flexibility in using SWAT for modeling hydrologic processes, however, there is a limitation in terms of high dependency on available data for calibration. The presence of errors to noise in the data warrants poor calibration of the models. The significance of the noise influence on the parameter estimation and model performance largely depends on the level and the nature of noise. There are two possible ways of tackling this situation: The first way is to have robust method for estimating the parameters values in the presence of noise, but this requires the knowledge about the structure and the variance of the noise apriori which is not possible in all cases. The second alternative is to remove the noise in the input variables (rainfall, temperature) using methods which does not require any apriori knowledge about the noise structure. Of these both ways, the latter method seems to be very useful in hydrological modeling context. In the past, Fourier transform was used as a denoising technique, but owing to its limitation for non-stationary time series, wavelet transform has been increasing used for denoising. This gives us an advantage to account for time variability of signals. With specific regards to denoising methods based on wavelets [4, 5] explored the denoising property of wavelets for maximization of ANN based model forecasting accuracy (in the context of flow forecasting). Model results show that networks trained with denoised data performed better than networks trained raw signals. There are numerous other studies (Kim and Valdes [5]; Partal and Kisi [6, 7]; Shiri and Kisi [8]; Kisi and Shiri [9]; Adamowski and Chan [10]; Belayneh and Adamowski [11]; Özger et al. [12]) which have shown that wavelet coupled with data-driven models provide a reliable framework for hydrological modeling. But there is no literature available till date which has explored the coupling of wavelets with physics-based model such SWAT. In this study, we approach the problem of improving hydrological model calibration by means of SWAT model coupled with wavelet denoising. For this purpose, the Wainganga basin is considered. Further, the impact of different wavelets (namely, Haar, db2, db3 and db5 and sym2) on removing the noise in the input signals and thereby aiding in calibration.

2 Soil Water Assessment Tool

The SWAT model works at Hydrologic Response Units (HRU) level. HRU are unique combination of land use, soil and slope and sub-basin. The model simulate the hydrological cycle using the water balance equation for each of the HRU. Detailed information about SWAT model can be found in [13, 14].

2.1 SWAT Model Calibration

Calibration of the SWAT model involves estimation of parameters which results agreeable simulation of the observable hydrologic processes such as runoff, evapotranspiration. The primary step in model calibration is the identification of the most sensitive parameters for a given catchment. Once the sensitive parameters are identified the model parameters are adjusted by either manually or using an automated programs. The manual calibration of models such as SWAT is quite tedious and time consuming. On the other hand automatic calibration is less subjective and capable of estimating the best parameters in less time. Recently, evolutionary algorithms are used extensively for estimating the parameters and have shown to be performing well on complex problems involving noisy functions.

There are many studies on calibration of SWAT model. Manual calibration is although widely used in calibration of SWAT models it is a tedious and daunting task that requires a trained hydrologist with a good knowledge of the watershed. On the contrary, automatic calibration is less subjective and it is capable of extensively searching sets of model parameters between their acceptable ranges in a very short period of time, increasing the likelihood of finding optimal parameters. The application of automatic methods in calibrating hydrologic models has become increasingly popular. Lately, evolutionary search techniques are being used in developing automatic routines for calibration due to the drawbacks associated with gradient-based algorithms. Especially, in cases where the calibration involves large number of parameters that have highly complex and nonlinear relationships with the model outputs, gradient-based search methods will have difficulties to converge to optimal sets of model parameters. Evolutionary algorithms, however, have been practically proved to perform well on complex optimization problems involving nonlinear, non-convex, and noisy functions [15].

The studies conducted on sensitivity of SWAT model parameters show that there are only 13 parameters which affect the quantity of runoff [16]. Out these, the parameters which are relevant for the case study under investigation is in Table 1 and have been chosen for the study. Table also shows the possible range for each of these parameters based primarily on ranges given by Neitsch et al. [13]. The parameter SURLAG controls the fraction of total water that is allowed to enter the stream on any particular day. It is significant in large basins where lag time is greater than 1 day. The ALPHA_BF is the index of how the groundwater flow

Table 1 SWAT parameters and range of the parameter values

Parameter name	Unit	Description	Range
CN_f	%		−0.25 to 0.15
ESCO	–	Soil evaporation compensation factor	0.001 to 1
SOL_AWC	%	Available water capacity	−0.3 to 2
SURLAG	day	Surface lag	0.1 to 12
SLOPE	%	Average slope steepness	−0.5 to 1
ALPHA_BF	day	The baseflow recession constant	0 to 1
OV_N	–	Manning's roughness coefficient for overland flow	0.1 to 0.3
GWDELAY	day	Groundwater delay time	0 to 500
GWREVAP	–	Groundwater revap. coefficient	0.02 to 0.2
GWQMN	mm	Threshold depth of water in shallow aquifer for return flow to occur	0 to 5000

responds to the recharge. GW_REVAP depends is a measure of the rate of the capillary rise of groundwater to the root zone and the rate of evapotranspiration. GW_DELAY is the time taken by water to cross the vadose zone and reach the aquifer. The ESCO is related to the how much evaporation is being contributed from different depths in the soil. Smaller the ESCO more evaporation starts taking place from deeper levels. SOL_AWC is the difference between field capacity and the wilting point. SLOPE is the average slope of HRUs in any sub-basin.

3 Wavelet Transform

Wavelets are orthogonal basis functions which are local in nature and can be used to represent a time series into a time-scale domain at different resolutions [17]. Using the multiresoultion ability, the noise features in a given signal can be removed to estimate the underlying original signal. In this study, the denoising capability of the wavelets is used to aid the calibration of the SWAT. For more details on mathematical background of wavelets refer [7].

3.1 Wavelet Denoising

The most commonly used method for denoising of signals is the threshold concept proposed by Donoho [18]. This method comprises of the following steps,

(a) Initially, a suitable mother wavelet and appropriate level of decomposition is chosen. Let the original signal $S(t)$ be decomposed into the M levels. This would results in M detailed components and one approximation component.
(b) Now, suitable threshold is selected and all $w_j(t)$ ($j = 1,2, \ldots, M$), are set to zero at each resolution level when MOD(w_j(t)) < threshold. The subscript j represents the jth resolution levels.
(c) Finally, the denoised time series is reconstructed from the approximation component and modified detailed signals $\hat{w}_j(t)$ at all resolution levels. The threshold has to be estimated in such a way that there is no significant information loss in the signal.

4 Case Study Description and Model Inputs

For demonstrating the proposed methodology a real time case study is considered. SWAT model is developed for the Wainganga basin in Godavari river basin. Wainganga is a tributary of the Godavari River. It originates in Seoni district in southern slopes of Satpura Range of Madhya Pradesh. After joining the Wardha, the united stream is known as the Pranahita. Pranahita ultimately falls into the Godavari River.

The study focuses on the hydrology of Wainganga basin of Godavari which occupies the area of 58,500 km^2 between latitudes 19°30′ N and 22°40′ N and longitudes 78°0′ E and 81°0′ E (Fig. 1).

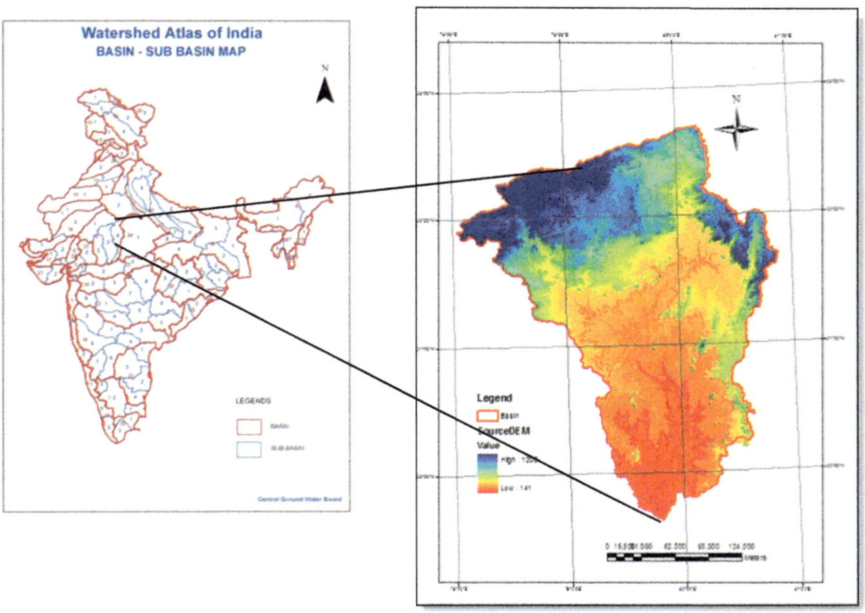

Fig. 1 Location of Wainganga basin

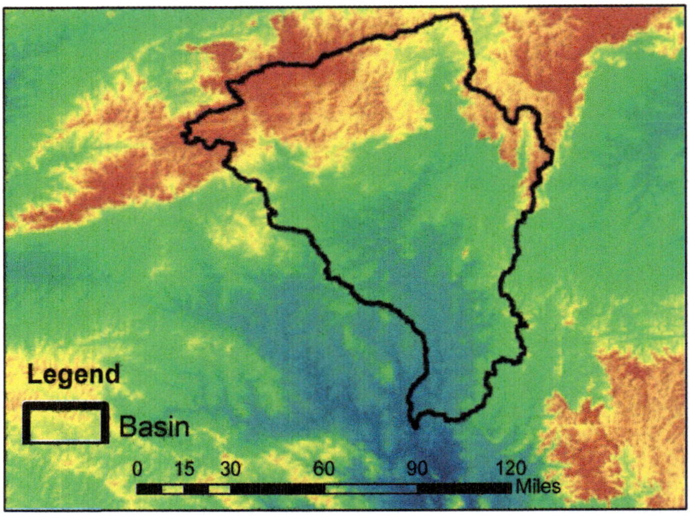

Fig. 2 Digital elevation map of Wainganga basin

In order to develop the model for hydrological simulation, SWAT uses a wide range of data which has to be provided as input. The following are the inputs which have been used in this case study. The DEM used in this study is extracted from the 90 m resolution SRTM dataset from CIAT and is downloaded from http://srtm.csi.cgiar.org has been used. The DEM for the study area is shown in Fig. 2. Land use and soil cover data which has to be added to the model has been obtained from the database of 'Global Land Cover Facility' which is available at: ftp.glcf.umd.edu/index.shtml. Moreover, the basin was further divided into sub-basins and Hydrological Response Units (HRU). Following factors formed the basis for this division:

1. specified threshold drainage area
2. homogeneity in land use
3. soil cover.

Daily temperature data from the period Jan 1, 1969 to Dec 31, 2005 of 14 stations was used taken. Daily rainfall data from 56 rain gauge station from the period Jan 1, 1969 to Dec 31, 2005 was used. Other meteorological data, as required by SWAT, namely solar radiation, wind speed, and relative humidity were estimated using the SWAT weather generator.

5 Model Calibration

5.1 SWAT Calibration Without Wavelet Denoising

The simulated discharge at the outlet of the basin was considered for calibration (Fig. 3). Calibration of SWAT was performed from the year 1991 to 1997 on monthly basis. The calibration of the model was done manually using Matlab interface. The parameters chosen for manual calibration were SURLAG, ALPHA_BF, GW_DELAY, GW_REVAP, GW_QMIN, ESCO, OV_N, SLOPE, SLSUBBSN, CN_F and SOL_AWC. All these parameters are assigned some default (initial) values by SWAT (Table 2). These parameters were altered one by one keeping all the other parameters unchanged. The same procedure is repeater till there is no change in model performance further. Figure 4 shows the discharge hydrograph of the measured and

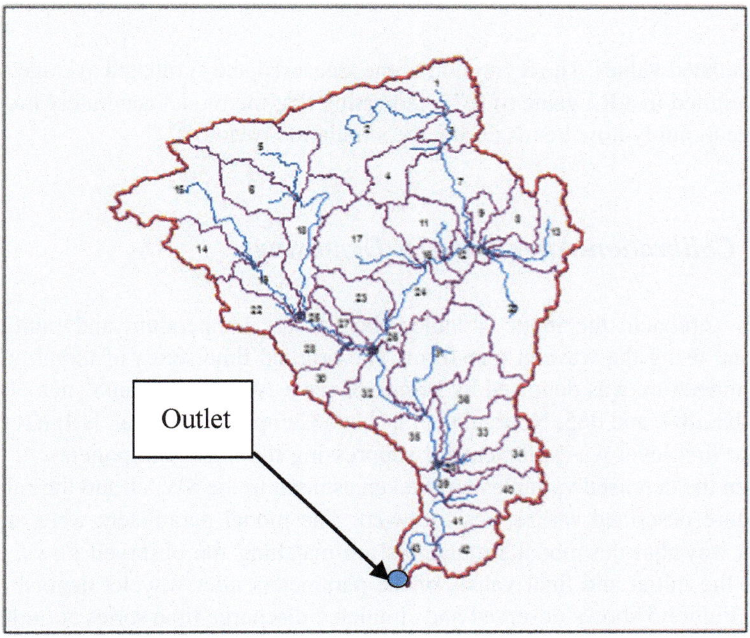

Fig. 3 Map showing the outlet for which calibration has been done

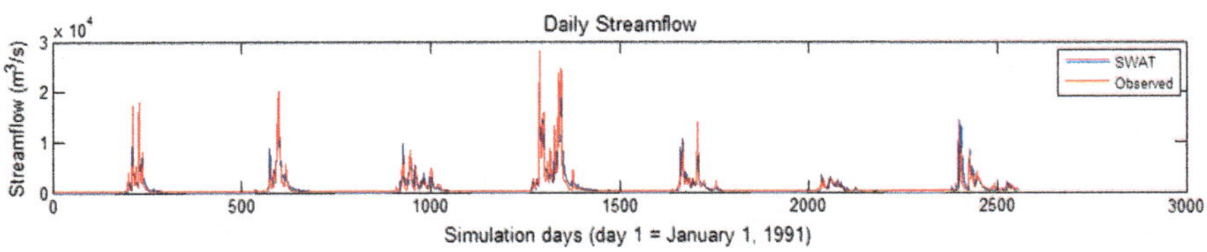

Fig. 4 Graph showing observed and simulated time series for original model from 1st Jan 1991 to 31st Dec 1997

Table 2 Initial and final values of the calibration parameters—manual calibration without denoising

Parameter	Unit	Initial values	Calibrated values
SURLAG	day	12	0.6
ALPHA_B	day	0.048	1
GW_DELA	day	31	40
GW_REVA	–	0.02	0.2
GWQMN	mm	0	10
ESCO	–	0.95	0.4
OV_N	–	0.15	0.293
SLOPE	%	0.582	0.582
SLSUBBS	%	0.11	0.11
CN_F	%	15	−0.07
SOL_AWC	%	1.764	1.764
		0.64	0.7341
		0.56	0.7332

the simulated values. The regression of the measured and simulated average monthly flow resulted in a R2 value of 0.73, indicating that the model accurately tracked the average monthly flow trends during the simulation period.

5.2 Calibration After Wavelet Denoising

In this approach, the input variables such as the temperature and rainfall were denoised using the wavelet transform. The original time series of monthly rainfall and temperature was denoised by means of six wavelet filter-banks, namely, Haar, db2, db3, db4, and db5. Nourani et al. [4] and Campisi-Pinto et al. [19] have shown that the first level was sufficient for suppressing the noise components.

Then the denoised variables were taken as input to the SWAT and the calibration procedure described earlier was followed. The model parameters were altered in such a way that the model simulations are matching the observed flows. Table 3 shows the initial and final values of the parameters after wavelet denoising using 'db2'. Figure 5 shows observed and simulated discharge time series at outlet of the basin from Jan 1, 1991 to Dec 31, 1997 for calibration after denoising. The better match of the calibrated time series and the observed time series can be observed. The R2 and NSC value for the model results after denoising was found to be 0.8699 and 0.8263. The NSC value has improved by 17%. This clearly shows that there is a drastic improvement in the model performance at the calibration stage. The results using the other wavelets is shown in Table 4. The comparison result revealed that the best mother wavelets which lead to better performance were mother wavelet db2 and mother wavelet db3 are appropriate for global denoising of rainfall and temperature series respectively.

Table 3 Initial and final values of the calibration parameters—manual calibration after wavelet denoising

Parameter	Unit	Default values	Calibrated values for denoised signal using 'db2'
SURLAG	day	12	0.6
ALPHA_B	day	0.048	0.9
GW_DELA	day	31	90
GW_REVA	–	0.02	0.2
GWQMN	mm	0	10
ESCO	–	0.95	0.4
OV_N	–	0.15	0.293
SLOPE	%	0.582	0.582
SLSUBBS	%	0.11	0.11
CN_F	%	0.15	−0.05
SOL_AWC	%	1.764	1.764
R2	–		0.8699
NSE	–		0.8263

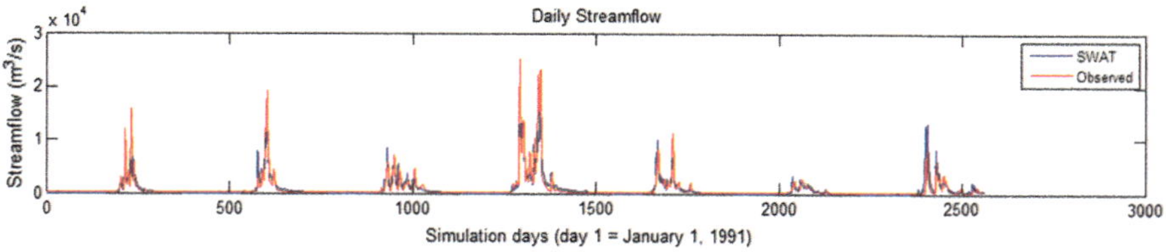

Fig. 5 Graph showing observed and simulated time series for scale 1 from 1st Jan 1991 to 31st Dec 1997

Table 4 Calibration results of SWAT model after wavelet denoising using various wavelets

Wavelet name	R2	NSC
Haar	0.8300	0.7912
db2	0.8699	0.826
db4	0.85	0.814
db5	0.82	0.76
sym4	0.824	0.768

6 Conclusions

An attempt was made in this paper to investigate the use of wavelet-based denoising technique as a preprocessing method for aiding calibration of SWAT model. The input time series such as rainfall and temperature was were denoised by the universal thresholding method. The denoised time series was used for calibration of the

SWAT model, the results reveal that the calibrated model with denoised inputs yielded better results in terms of NSC than the model with raw inputs. Thus, wavelet-based denoising can be reliably used to improve the calibration and aid in modeling. The sensitivity analysis of different wavelets reveal that the db3 and db2 wavelets perform better than other wavelets in denoising and thereby the model calibration results.

References

1. Arnold, J.G., Srinivasan, R., Muttiah, R.S., Williams, J.R.: Large area hydrologic modeling and assessment part I: model development1. Wiley Online Library (1998)
2. Srinivasan, R., Ramanarayanan, T.S., Arnold, J.G., Bednarz, S.T.: Large area hydrologic modeling and assessment part II: model application1. Wiley Online Library (1998)
3. Zhang, X., Srinivasan, R., Van Liew, M.: Multi-site calibration of the SWAT model for hydrologic modeling. Trans. ASABE **51**(6), 2039–2049 (2008)
4. Nourani, V., Rahimi, A.Y., Nejad, F.H.: Conjunction of ANN and threshold based wavelet denoising approach for forecasting suspended sediment load. Int. J. Manage. Inf. Technol. **3** (1), 09–25 (2013)
5. Kim, T.-W., Valdes, J.B.: Nonlinear model for drought forecasting based on a conjunction of wavelet transforms and neural networks. J. Hydrol. Eng. **8**(6), 319–328 (2003)
6. Partal, T., Kisi, M.: Long-term trend analysis using discrete wavelet components of annual precipitations measurements in Marmara region (Turkey). Phys. Chem. Earth Parts A/B/C **31** (18), 1189–1200 (2006)
7. Partal, T., Kisi. Wavelet and neuro-fuzzy conjunction model for precipitation forecasting. J. Hydrol. **342**(1), 199–212 (2007)
8. Shiri, J., Kisi, O.: Short-term and long-term streamflow forecasting using a wavelet and neuro-fuzzy conjunction model. J. Hydrol. **394**(3), 486–493 (2010)
9. Kisi, O., Shiri, J.: Precipitation forecasting using wavelet-genetic programming and wavelet-neuro-fuzzy conjunction models. Water Res. Manage. **25**(13), 3135–3152 (2011)
10. Adamowski, J., Chan, H.F.: A wavelet neural network conjunction model for groundwater level forecasting. J. Hydrol. **407**(1), 28–40 (2011)
11. Belayneh, A., Adamowski, J.: Standard precipitation index drought forecasting using neural networks, wavelet neural networks, and support vector regression. Appl. Comput. Intell. Soft Comput. **6** (2012)
12. Özger, M., Mishra, A.K., Singh, V.P.: Long lead time drought forecasting using a wavelet and fuzzy logic combination model: a case study in Texas. J. Hydrometeorol. **13**(1), 284–297 (2011)
13. Neitsch, S.L., Arnold, J.G., Kiniry, J.R., Srinivasan, R., Williams, J.R.: Soil and Water Assessment Tool User's Manual Version 2000. GSWRL report 02-02, BRC report 02-06. Texas Water Resources Institute TR-192: College Station, TX. 438 (2002)
14. Gassman, P.W., Reyes, M.R., Green, C.H., Arnold, J.G.: The soil and water assessment tool: historical development, applications, and future research directions. Center for Agricultural and Rural Development, Iowa State University (2007)
15. Schwefel, H.-P., Rudolph, G.: Contemporary Evolution Strategies. Springer (1995)
16. Santhi, C., et al.: Validation of the swat model on a large Rwer basin with point and nonpoint sources. Wiley Online Library (2001)
17. Grossmann, A., Morlet, J.: Decomposition of Hardy functions into square integrable wavelets of constant shape. SIAM J. Math. Anal. **15**(4), 723–736 (1984)

18. Donoho, D.L.: De-noising by soft-thresholding. IEEE Trans. Inf. Theory **41**(3), 613–627 (1995)
19. Campisi-Pinto, S., Adamowski, J., Oron, G.: Forecasting urban water demand via wavelet-denoising and neural network models. Case study: city of Syracuse, Italy. Water Res. Manage. **26**(12), 3539–3558 (2012)

Hydrogeophysics and Numerical Solute Transport Modelling Techniques for Environmental Impact Assessment

L. Surinaidu, V. V. S. Gurunadha Rao and Y. R. Satyaji Rao

Abstract The present study is carried out to assess the impact of upcoming Aluminium industry impact on groundwater quality. Hydrogeological, electrical resistivity tomography and numerical solute transport modelling techniques are deployed to understand aquifer nature and anticipated transport of contaminants impact on groundwater quality from the proposed red mud pond in the Visakhapatnam, Andhra Pradesh. The hydrogeological results revealed that the area possesses low infiltration rate that is range from 1.26 to 5.68 cm/h with moderated hydraulic conductivity range from 3.5 to 5.6 m/day. The groundwater quality is in good health within the range of drinking water standards of WHO/BIS. Geophysical data analysis revealed that weathered zone thickness range from 10 to 25 m from the surface to subsurface then it is underlain by fissured/fracture layer of 5–10 m thick followed by hard rock. The numerical solute transport modelling predicted that there would not be any significant increase in TDS concentration for next 50 years outside of the upcoming red mud pond under the observed hydrogeological conditions of the year 2009 and the range of assumed dynamic loading of pollution in terms of TDS. The techniques can be utilized to select the suitable site for waste disposals and impact assessment of existing contaminant source on groundwater.

Keywords ERT · Hydrogeophysics · Water quality · Infiltration and solute transport

L. Surinaidu (✉) · V. V. S. Gurunadha Rao
CSIR-National Geophysical Research Institute (NGRI), Hyderabad, Telangana, India
e-mail: L.Surinaidu@ngri.res.in

Y. R. Satyaji Rao
National Institute of Hydrology (NIH), Kakinada, Andhra Pradesh, India

M. Rathinasamy et al. (eds.), *Water Resources and Environmental Engineering I*,
https://doi.org/10.1007/978-981-13-2044-6_14

1 Introduction

In the recent year's groundwater aquifers are facing serious water quality degradation and water level depletion problems due to intensive irrigation, expansion of industrialization and urbanization [1]. Industrial contamination is one of a major threat to fresh water aquifers due to the unplanned disposal of industrial wastes. The proper planning of site selection and preparation for waste disposal with proper engineering design can protect the aquifers from any kind of contamination. Hydrogeophysical and numerical solute transport modelling tools are promising techniques to understand subsurface geometry and solute movement in the subsurface for subsequent planning and management. Many researchers have been applied geophysical techniques for the proper site selection and environment impact assessment [2] numerical flow and solute transport modelling is applied for evaluation and predicting existing and future groundwater levels and contamination from various sources [3–5].

In the present study an Alumina refinery is going set up in Visakhapatnam, Andhra Pradesh for the production of Alumina. During its production red mud will be generated after extraction of Alumina from bauxite ore. The disposal of the red mud needs proper site selection and evaluation. The present study is carried out with the aim of assessing likely impact of upcoming red mud pond on groundwater quality using hydrogeological, geophysical and numerical solute transport modelling for the entire watershed that includes proposed red mud pond covering 120 km^2 (Fig. 1). The proposed areal extent of red mud is about 50 ha. The area forms a part of Eastern Ghat Mobile belt exposing all the characteristic litho units of the Eastern Ghat Super group such as the Khondalite, Charnockite and migmatite. The area receives an average annual rainfall of 1050 mm/year.

2 Data and Methodology

2.1 Hydrological Investigations

Infiltration test area carried out using double ring in filter metre. Single well pump tests are carried out at three locations by pumping the groundwater from the well for 4 h with measured constant discharge rate. The recovery of water levels are measured after stopping pumping in the well until it reaches its previous water levels. The pump test data is interpreted using Aquifer Test software [6]. The detailed groundwater monitoring is carried out at 19 locations by measuring groundwater level depths by standard water level indicator during March, June and November 2009 and topography survey is carried out using total station survey method. Groundwater samples have been collected from all observation wells and analysed for major ions by following standard water quality analysis methodology of APHA, 2005.

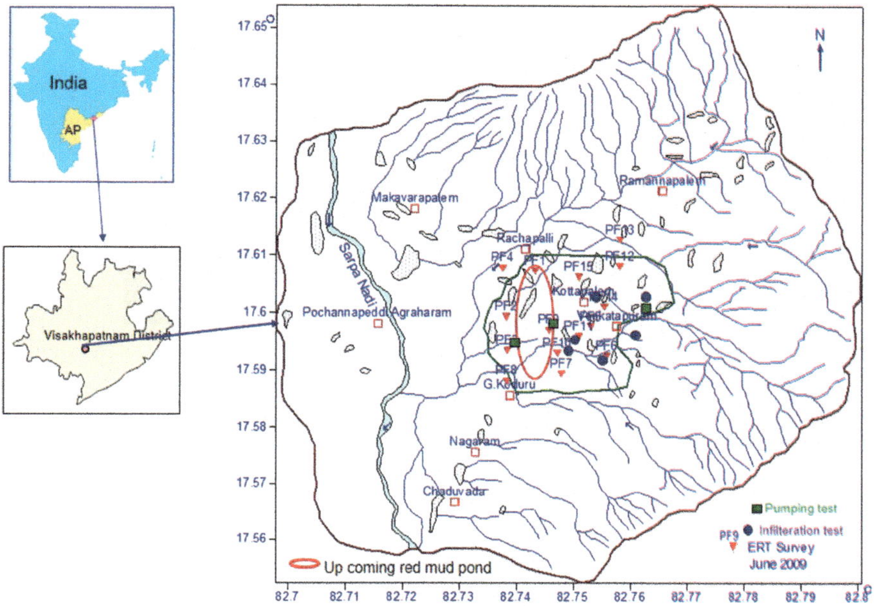

Fig. 1 Location of ERT, infiltration and pump tests in the study area, Visakhapatnam, Andhra Pradesh

2.2 *Geophysical Investigations*

Geophysical investigations in the form of electrical resistivity tomography (ERT) have been carried out at 15 locations in the study area (Fig. 1). The apparent resistivity data was collected using SYSCAL Pro resistivity metre with 96 electrodes using 5 m electrode spacing with Wenner–Schlumberger, Wenner and dipole–dipole configurations to delineate aquifer geometry and to identify the potential groundwater zones. The collected apparent resistivity values are then converted into true resistivity using RES2D.INV inversion programme [7]. The data inversion is calculated with the least squares inversion method [8] 2D resistivity data is interpreted using 2D-forward modelling software. Apparent resistivity values are calculated using finite-difference method [9].

2.3 *Groundwater Flow and Solute Transport Modelling*

The numerical groundwater flow and solute transport model is constructed using USGS MODFLOW and MT3DMS code with Visual Modflow as a graphical interface for watershed covering Aluminium industry spreads over 120 km^2 that was conceptualized as a two layered weathered and fractured Khondalitic aquifer

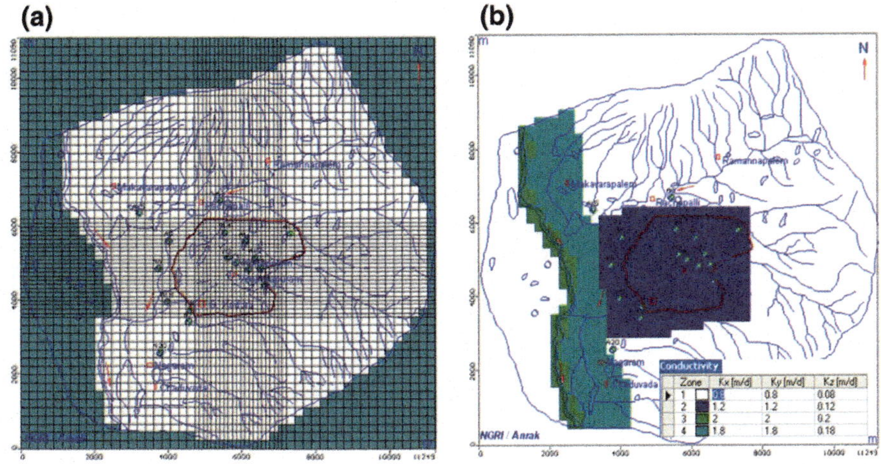

Fig. 2 Grid map of the groundwater flow and solute transport modelling (**a**), Hydraulic conductivity distribution in the model (**b**)

system. The model domain has 62 rows × 61 columns of variable cell sizes viz., 225 m × 225 m and 122.5 m × 122.5 m (Fig. 2a). The fine grid spacing has been chosen around the proposed Red Mud Ponds. The total thickness of simulated aquifer is 55 m with an average thickness of the weathered zone is 35 m. The simulated permeability of the saturated Khondalitic formations has been varying from 0.8 to 2.0 m/day in the groundwater model (Fig. 2b). Permeability of weathered rock underlying the Red mud ponds have been simulated as 1.2 m/day, which has been assumed to be slightly lower compared to the estimated value from the pumping test. The permeability value has been 2 m/day due to larger weathering of colluvial formations around the streams draining towards the Sarpanadi. No significant lineament has been found in the proposed red mud ponds area, which indicates that the groundwater flow through fracture/joint formations is negligible in the red mud ponds area. Hence homogeneous permeability distribution has been assigned in both *X* and *Y* directions for simulation of lateral groundwater movement in the area. The permeability in the vertical direction has been assumed to be one tenth of the horizontal permeability.

A Constant head boundary condition has been simulated at the outlet of the Sarpanadi coined with a groundwater head of 30 m (amsl) to allow lateral groundwater outflow from the south boundary adjacent to Sarpanadi. The Sarpanadi has been simulated with a river boundary condition along its course varying from 46 m in the north to 30 m (amsl) at the outlet in the south (Fig. 3a). The hydraulic conductance between aquifer and stream has been assumed as 50 m^2/day, which will be controlling the stream aquifer interaction. The average groundwater recharge would be very low and has been assumed to be varying from 65 mm/year in the uplands to 40 mm/year in the valley parts near the Sarpanadi (Fig. 3b). The recharge is assumed to be at 45 mm/year in the proposed red mud ponds area.

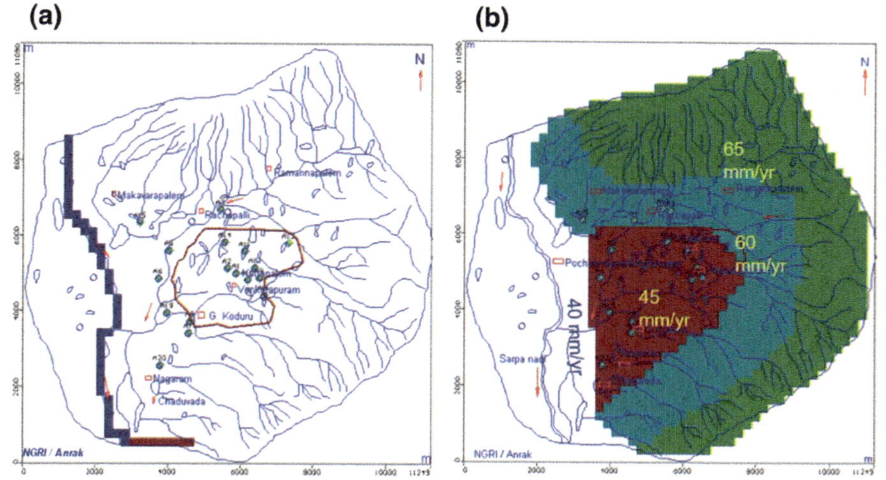

Fig. 3 River (blue colour) and constant head (maroon colour) condition (**a**), Recharge distribution in the study area (**b**)

As regards to groundwater withdrawal, only few irrigation wells are withdrawing groundwater for agriculture with an average pumping rate of 100 m^3/day. The same has been assigned in model using well package (refer Fig. 6b).

3 Results and Discussion

3.1 Hydrogeology

The topography elevation in the area is range from 62.1 to 40 m (amsl) from East to West towards the Sarpanadi (Fig. 4a). The proposed red mud pond stacking area is located at the elevation of 48–52 m (amsl). The depth to groundwater level range from 2.7 to 10.6 m (bgl) during pre-monsoon (March 2009) where is in the post-monsoon it is range from 1.5 to 3.5 m (November 2009) (Table 1). The depth to groundwater level is shown only 1.8 m (bgl) in the dug well near Rachapalli village close to the proposed red mud ponds. The average depth to groundwater level is about 5 m in the red mud ponds area between Rachapalli and G. Koduru during pre-monsoon. The regional groundwater elevations range from 52 to 34 m (amsl) during March 2009. The groundwater flow direction is from the Aluminium factory site towards the Sarpanadi could be seen with steep gradients around Koduru village (Fig. 4b). Similar trends have been noticed with regard to depth to groundwater level and regional groundwater flow during post-monsoon of 2009. The in situ infiltration rates are found to be varying from 1.26 to 5.68 cm/h. The infiltration rate in the red mud pond area has been found slightly high. The higher values may be attributed to present land use of the red mud pond area for

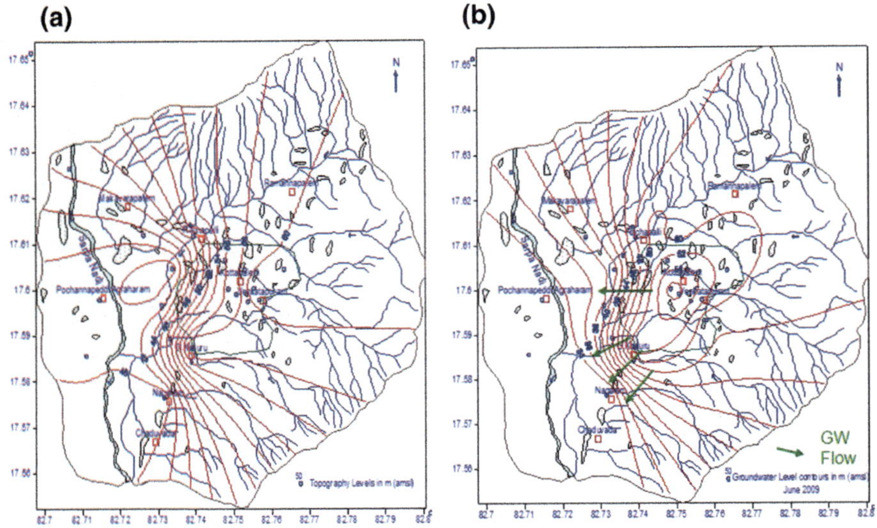

Fig. 4 Topography contours in m (amsl) (**a**) and Groundwater level contours in m (amsl) (**b**) in the study area

agriculture. Once the agriculture operations cease, the top soil zone may become harder as there would be no tillage. Hence the present infiltration rates estimated would represent the maximum possible infiltration rate in the area (Table 2). The estimated average hydraulic conductivity of the area is found to be varying from 3.5 to 5.6 m/day. The average hydraulic conductivity of 4.5 m/day has been worked out for the red mud ponds area (Table 3).

3.2 Groundwater Quality

Groundwater quality analyses results of major ions during pre-monsoon (March) and post-monsoon (November) of 2009 is presented in Table 4. The statistical analysis of major ions in groundwater also presented including 25, 50, and 75% quintiles indicate that most of the samples show good water quality (Table 4). The pH value is varying from 7.1 to 8.5 and 7.5–8.5 during pre-monsoon and post-monsoon 2009 respectively. Total dissolved solids (TDS) concentration in groundwater has been found to be ranging from 700 to 1340 mg/L in the watershed during pre-monsoon. The average TDS concentration in the proposed red mud ponds area is around 800 mg/L. Post-monsoon groundwater quality showed slight reduction in TDS values that may be due to dilution is taking places from natural groundwater recharge during monsoon rainfall. The back ground chloride concentration in groundwater is about 200 mg/L during pre monsoon and post-monsoon periods. The fluoride concentration in groundwater has been found range from 0.75 to 1.52 mg/L during

Table 1 Groundwater levels in m (amsl) in the study area, Visakhapatnam district, Andhra Pradesh

S. No.	Well type	Longitude °E	Latitude °N	Depth to groundwater level in m (bgl)			Reduced level in m (amsl)	Groundwater level in m (amsl)		
				March 2009	June 2009	November 2009		March 2009	June 2009	November 2009
A1	DW	82.75072	17.59903	4.42			56.5	52.08		
A2	HP	82.74864	17.60019				56.8			
A3	DW	82.74864	17.61403	5.4	5.6	2.3	53.26	47.86	47.66	50.96
A4	SW	82.73944	17.61164				48.23			
A5	DW	82.73381	17.60456			1.52	38.9			37.38
A6	DW	82.73097	17.59764	2.73	2.86	0.9	40.97	37.64	37.51	39.47
A7	HP	82.73875	17.58431	3.5	3.8	1.8				
A8	HP	82.73897	17.58717	5.4	5.9	3.2	60.34	54.44	53.94	56.64
A9	DW	82.75675	17.59778	4.87		3.5	60.47	54.6		55.97
A10	HP	82.75378	17.59744							
A11	DW	82.75297	17.60442	2.95	2.98	1.23	58.46	55.51	55.48	57.23
A12	DW	82.75542	17.60033	6.68	6.75	4.72	62.11	54.43	54.36	56.39
A13	DW	82.75667	17.61869	7.58	7.59	3.35	60.21	51.63	51.62	55.86
A14	DW	82.74781	17.60653	5.5	5.3	1.85	56.04	50.54	50.74	54.19
A15	HP	82.72622	17.61183	3.7	3.2	2.7				
A16	HP	82.71458	17.59783							
A17	HP	82.71114	17.58556							
A18	SW	82.70672	17.62644							
A19	HP	82.73314	17.58931							
A20	HP	82.73153	17.57644	8.2	7.9	4.52	38.35	29.65	29.95	33.33
A21	BW	82.75792	17.59322	10.6	11.2	6.47	59.75	49.15	48.55	53.28

Table 2 In situ infiltration rate in the study area, Visakhapatnam district, Andhra Pradesh

Test no.	Location	Infiltration rate (cm/h)
1	Inside smelter area near to DW No. A12	1.52
2	Inside proposed refinery plant area	1.26
3	Inside proposed red mud pond area near Rachapalli village	4.84
4	Inside middle of proposed red mud pond area	5.42
5	Inside proposed red mud pond area, Near to G. Koduru village	5.68

Table 3 Estimated aquifer parameters in the study area, Visakhapatnam district, A.P.

	Theis Method	Neuman method	Cooper & Jacob
PW1 (Rachapalli)			
Transmissivity (m^2/day)	28.2	25.6	17.6
Hydraulic conductivity (m/day)	5.6	5.13	3.52
PW2 (Dharmavaram)			
Transmissivity (m^2/day)	20.4	17.0	18.7
Hydraulic conductivity (m/day)	17.3	14.4	16.0
PW3 (Koduru)			
Transmissivity (m^2/day)	15.3	19.9	19.8
Hydraulic conductivity (m/day)	4.05	5.26	5.22

pre-monsoon and 0.85–1.42 during post-monsoon. The sulphate concentration is found to be <100 mg/L during both monitoring periods. As there is only sparse vegetation in the area, all samples have reported low the nitrate as *N* values. The sodium concentration in groundwater is found ranging from 111 to 330 mg/L during pre-monsoon and post-monsoon periods. The background sodium concentration has been around 200 mg/L in the proposed red mud ponds area. The calcium concentration in groundwater has been found <50 mg/L. The potassium and magnesium concentration have been found to be very low in groundwater. In general the water quality in the area is found to be potable and safe within the WHO/BIS drinking water standards [10, 11].

3.3 Geophysical Investigations

A total of 15 ERTs have been carried out at 15 locations in the sturdy area (refer Fig. 1 for locations).The representative resistivity cross sections along the profiles (PF) at six locations have been presented and discussed in the present paper (Fig. 3). The depth of investigation is limited to a maximum of 65 m in the

Table 4 Statistical Analysis of groundwater quality data in the study area, Visakhapatnam District, A.P, 2009

	March 2009						November 2009					
Parameter	Min	Max	Average	Q25	Q50	Q75	Min	Max	Average	Q25	Q50	Q75
pH	7.1	8.7	7.61				7.3	8.5	7.75			
TDS	157	1523	916	607	861	1214	204	1358	817	651	819	985
Cl	25	350	163	66	120	248	20	310	131	70	110	170
SO_4	6	110	47	22	37	62	6	103	43	25.5	38	46
NO_3–NO_3	1.80	310	51.38	5.40	13.30	37.50	1.80	239	34.32	3.20	9.90	47.70
F	0.3	2.89	1.27	1	1	2	0.32	2.53	1.26	0.9	1.3	1.45
Na	13	338	187	109	196	243	23	375	192	131	191	219
K	2	164	16	5	5	8	5	50	10	5	5	7
Mg	2	117	47	22	42	61	10	73	42	29	39	58
Ca	24	80	43	32	40	48	8	28	14	8	16	16

All values are in mg/L except pH

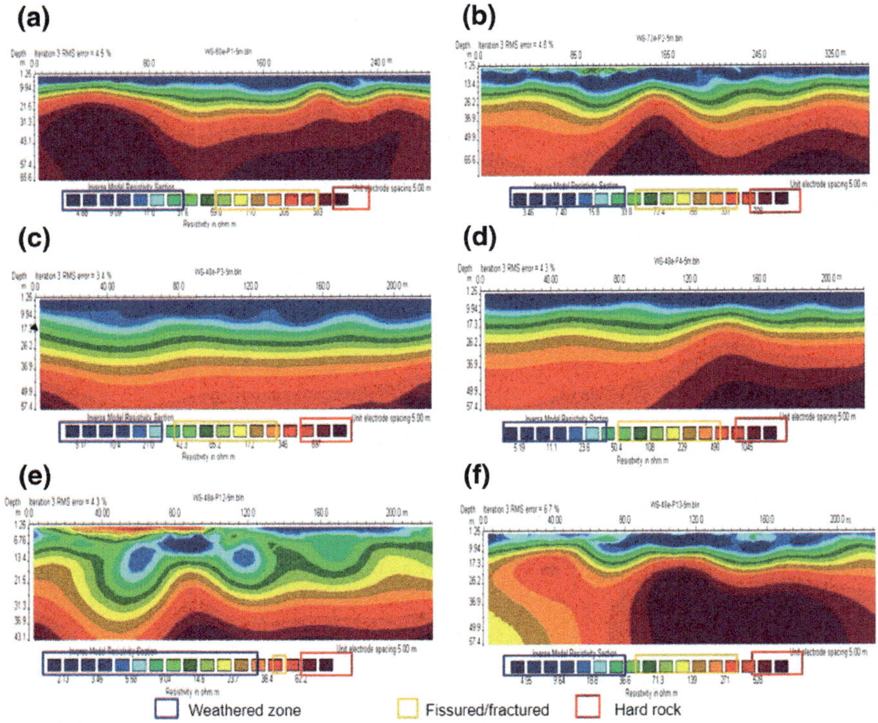

Fig. 5 Electrical resistivity cross sections obtained from ERT

subsurface in all the ERT cross sections. In general, all ERT images shown that the weathered zone is extending from 10 to 25 m depth followed by a fracture zone of 5–10 m thickness and it is underlaid by basement rock.

PF1 has been carried out in the proposed red mud pond area in Rachapalli village. The profile is laid approximately in the North–South direction. The ERT image indicates a low resistivity zone up to a depth of 5 m underlain by weathered zone extending up to a depth of about 15 m. A fractured zone of 5 m thickness could be seen below the weathered zone. It is underlain by hard compact rocks up to a depth of ~65 m with elevated resistivity values (Fig. 5a).

PF2 is laid in NE–SW direction and the ERT image shows the thickness of formations in the central part of the proposed red mud pond area. The inverse model resistivity cross section indicates a low resistivity layer up to a depth of about 10 m. The layer is underlain by a weathered zone extending up to 25 m depth. Hard compact rocks of Charnockite formations have been encountered from 25 to 65 m depth with high resistivity values >700 ohm m (Fig. 5b).

Profile No. 3 (PF3) The ERT transact was carried out in the southern part of proposed Red mud pond area. The profile was laid in the NE–SW direction. The ERT image represents a low resistivity layer up to a depth of ~11 m underlain by a weathered zone up to depth of about 25 m. Below the weathered zone, a fractured zone of ~10 m thickness extends up to 35 m depth. The presence of Hard Charnockite rocks has been inferred with high resistivity values underlying the fracture zone (Fig. 5c).

Profile No. 4 (PF4) The ERT profile has been carried out in the paddy field adjacent to the proposed Red mud pond area in Rachapalli village. The profile was laid in the North–South direction parallel to the PF1. The image represents a low resistivity layer up to a depth of about 18 m underlain with compact soft rock and hard rocks with varying resistivity values and extends to a depth of ~57 m (Fig. 5d).

Profile No. 12 (PF12) Electrical Resistivity Tomography survey was also carried out inside proposed Surface water Pond area and the profile was laid in N–S direction. The inverse model resistivity image indicated a low resistivity zone up to a depth of 13 m in the top followed by a thick-weathered zone extending up to 43 m depth. Some engineering interventions like HDPE liner placement at the bottom of the surface water reservoir suggested for arresting seepage losses from the proposed surface water storage reservoir (Fig. 5e).

Profile No. 13 (PF13) Electrical Resistivity Tomography was also carried out inside proposed Surface water Pond area, about 150 m away from PF. No. 12 and the profile was laid in N–S direction. The inverse model resistivity image indicated a low resistivity zone up to a depth of 5 m, and the layer is underlain by a weathered zone up to a depth of 10 m, which is followed by the fractured zone up to a depth of 18 m. Hard compacted rocks have been encountered in the middle of the inverse resistivity model transact, which extends towards the south (Fig. 5f). In the area low resistivity values at the shallow depth may be indicating saturated conditions and also shallow groundwater table. To avoid any infiltration of pollutants, proper HDPE liners with proper engineering design may use in the bottom of any contaminant disposals in the area.

3.4 Groundwater Flow and Solute Transport Modelling

The computed versus observed hydraulic heads at 19 observation wells has been found matching closely particularly in the proposed Red Mud Pond area (Fig. 6a). The computed groundwater velocity vectors indicate predominant flow towards the Sarpanadi and also to the south with an average groundwater velocity of ~15 m/year. A longitudinal dispersivity of 40 m has been assumed and the ratio of horizontal to longitudinal dispersivity of 0.1 and vertical to longitudinal dispersivity of 0.01 has been used as dispersion parameters in the mass transport simulation. Conservatively, the back ground initial TDS concentration in groundwater hasbeen assumed as 700 mg/L. The expected dynamic loading of leachate from the proposed red mud Ponds area has been simulated by assigning TDS concentrations as

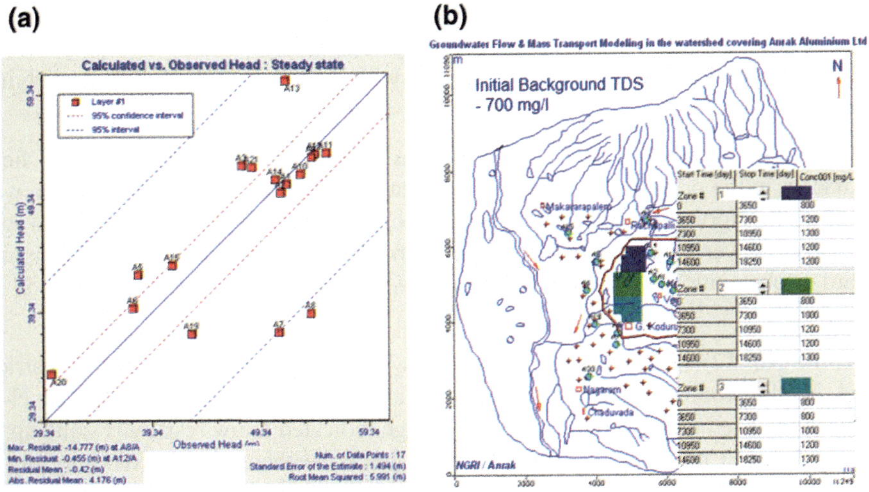

Fig. 6 Calculated versus observed groundwater levels m (amsl) (**a**) and dynamic TDS (mg/L) loading at proposed red mud ponds for the next 50 years in the MT3DMS (**b**)

per design of red mud stacking scenario in the ponds during next 50 years period (Fig. 6b). The maximum leachate of TDS concentration from the red mud ponds that may leach to the groundwater table would be 1300 mg/L at peak loading has been assumed and later starts decreasing as stacking of red mud shifts to adjacent ponds over a period of time. The predicted lateral migration of TDS concentration from the proposed red mud ponds from the Mass transport model does not indicate that the TDS plume concentration >1200 mg/L reaching out of the assigned Anrak land during next 50 years (Fig. 7a–d).

Lateral migration of the leachate plume is limited to red mud ponds area only and the highest TDS concentration attained would be 1000 mg/L during next 10 and 20 years. Further the leachate plume spreads in the down gradient and the TDS plume concentration does not exceed 1100 mg/L during next 30 and 40 years. Attaining highest TDS concentration of 1200 mg/L within the red mud ponds could be seen only after 50 years.

As regards to the movement of leachate plume along the vertical direction, the cross section view of the predicted TDS plume along Column No. 23 did not indicate faster movement of TDS plume at depth during next 50 years (Fig. 8a, b). The leachate plume from the red mud Pond with TDS concentration of about 800 mg/L would spreads out to the bottom of the first layer (weathered zone) only after 20 years of red mud loading. The predictions made from the mass transport model does not indicate TDS plume migrating out of the red mud pond area during next 50 years with elevated concentrations. To monitor the movement of pollution over the time on regular basis from the red mud pond, shallow piezometers up to 10–20 m depth should be drilled at 4–6 locations close to the boundary of red mud ponds.

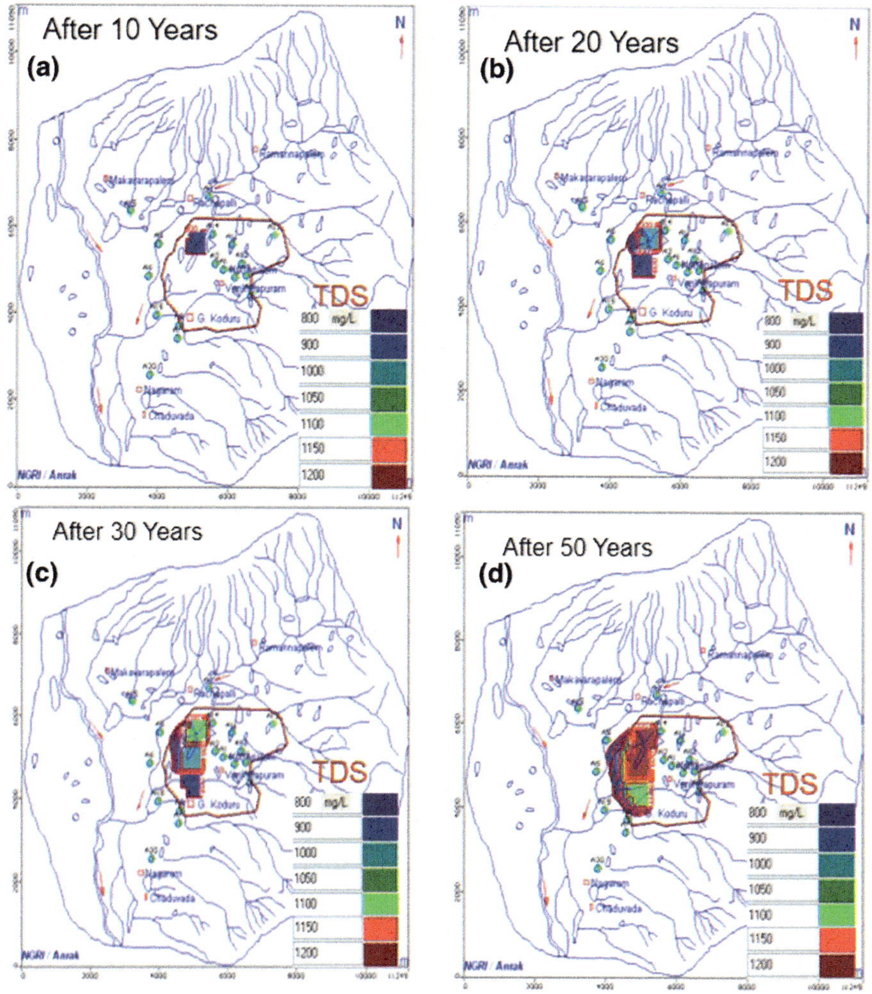

Fig. 7 Lateral transport of TDS concentration from proposed red mud pond for different years

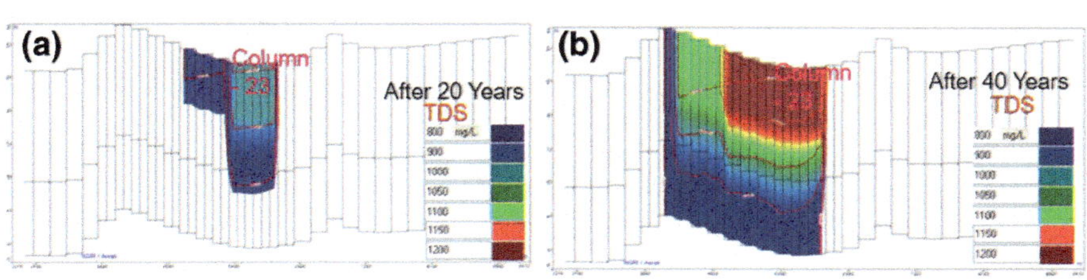

Fig. 8 Vertical transport of TDS after 20 and 40 years from the proposed red mud ponds

4 Conclusions

The study area is occupied by moderately resistive Khondalitic formation with poor recharges characteristics. The groundwater recharge potential is very low as major part of the area contributes for surface runoff. It is suggested to divert stream channels passing close to the red mud ponds and raise the bunds of red mud ponds to arrest entry of surface runoff. Groundwater fluctuations from pre-monsoon to post-monsoon is varying from 2 to 5 m with average groundwater level rise of <1.5 m after post-monsoon. The average in situ infiltration rate is 4 cm/h in the area, which may be further reduced due to anticipated ceasing of agriculture operations and irrigation practice. It is suggested that some engineering intervention in the red mud pond design is required to isolate the ground surface as well as shallow groundwater table in the area. The groundwater velocities computed from the groundwater flow model in the watershed covering the red mud pond area is $\sim$15 m/year. The computed TDS plume from numerical solute transport modelling indicated that there would not be significant increase in TDS concentration outside of the proposed red mud pond area for the next 50 year from 2009. The approved design of HDPE liners of appropriate thickness may be incorporated at the bottom of the red mud stack as per CPCB guide lines to further minimize leachate movement into the groundwater regime. Drilling of 4–6 numbers of shallow piezometers up to 10–20 m depth at the boundary of red mud ponds has been recommended for monitoring of groundwater level and groundwater quality on a regular basis.

Acknowledgements Authors would like thank Director of CSIR-National Geophysical Research Institute (NGRI), Hyderabad for his encouragement to publish this paper. This project is funded by ANRAK ALLUMINIUM, Pvt. Ltd. Vishakhapatnam. Authors also would like to thank Mr. Tamma Rao for his help in carrying out field work.

References

1. Howard, K.W.F., Gelo, K.K.: Intensive Groundwater use in Urban Areas: The Case of Megacities (2002)
2. Pujari, P.R., Padmakar, C., Surinaidu, L., Vaijnath, V.U., Kachawe, B., Rao, V.G., Labhasetwar, P.K.: Integrated hydro chemical and geophysical studies for assessment of groundwater pollution in basaltic settings in Central India. Environ. Monit. Assess. **184**(5), 2921–2937 (2012). https://doi.org/10.1007/s10661-011-2160-1
3. Surinaidu, L., Rao, V.V.S.G., Prasad, P.R., Sarma, V.S.: Use of geophysical and hydrochemical tools to investigate seawater intrusion in coastal alluvial aquifer, Andhra Pradesh, India. In: Groundwater in the Coastal Zones of Asia-Pacific, vol. **7**, pp. 49–65. Springer publications (2013)
4. Surinaidu, L., Rao, V.V.S.G., Mahesh, J., Prasad, P.R., TammaRao, G., Sarma, V.S.: Assessment of possibility of saltwater intrusion in the central Godavari delta region, Southern India. Reg. Environ. Change **15**(5), 907–918 (2014). https://doi.org/10.1007/s10113-014-0678-9

5. Surinaidu, L., Mutuwatte, L., Amarasinghe, U.A., Jain, S.K., Ghosh, N.C., Kumar, S., Singh, S.: Reviving the Ganges water machine: accelerating surface water and groundwater interactions in the Ramganga Sub-basin. J. Hydrol. **540**, 207–219 (2016). https://doi.org/10.1016/j.jhydrol.2016.06.025
6. Waterloo hydrologic (2016). https://www.waterloohydrogeologic.com/wp-content/uploads/PDFs/aquifertest/AquiferTest2016_UsersManual_web.pdf
7. Loke, M.H.: Electrical Imaging Surveys for Environmental and Engineering Studies. Unpublished report, pp. 1–8 (1997)
8. Dahlin, T.: 2D resistivity surveying for environmental and engineering applications. FirstBreak **14**, 275–284 (1996)
9. Dey, A., Morrison, H.F.: Resistivity modelling for arbitrary shaped two-dimensional structures. Geophys. Prospect. **27**, 1020–1036 (1979)
10. Bureau of Indian standards (BIS). ManakBbhavan, Bbahadur shah zafarmarg, Publication unig, New Delhi (2012). http://cgwb.gov.in/Documents/WQ-standards.pdf
11. World Health Organization (WHO): Guidelines for Drinking-Water Quality [Electronic Resource]: Incorporating First Addendum, vol. 1, Recommendations, 3rd edn. (2012) http://www.who.int/water_sanitation_health/dwq/gdwq0506.pdf

Poor Storm Water Drainage and Stripping on a Stretch of NH-1

Gourav Goel and S. N. Sachdeva

Abstract Improper drainage facilities on a road can cause failure of its pavement. Accumulation of water on roads during monsoon/rains causes inconvenient and retarded traffic movement. It can result in loss of adhesive bond between aggregate and bitumen on the road surface resulting into stripping and threatens the durability of roads. Proper drainage system with adequate section of storm water drains can ensure the safety against above aspects. In this paper effect of poor drainage system on road pavement and adequacy of storm water drain on a stretch of NH-1 has been studied. It is found that the size/capacity of drain is sufficient but it is not able to drain the water due to poor maintenance of drains and adverse slope of the ground between drain and the road. It is also found that stripping value of bituminous mix increases with the increase in immersion time of aggregate in water.

Keywords Improper drainage · Highway pavement · Stripping
Storm water drain

1 Introduction

Roads are lifeline of any country and proper drainage system on roads is one of the basic requirements of a road project. Poorly drained roads not only reduce traffic capacity of the road but also affects road pavement. Moisture presence of road can lead to removal of bitumen film from the aggregate surface resulting into stripping phenomenon. Inadequate drainage leads to stripping of asphalt in flexible pavements. Pavement service life can be extended by 50% if water is drained off from the road effectively [1]. Poor drainage system also increases maintenance cost of the roads. It is necessary that due priority is given to the drainage of roads and

---

G. Goel (✉) · S. N. Sachdeva
Department of Civil Engineering, National Institute of Technology,
Kurukshetra 136119, Haryana, India
e-mail: gouravgoelshinu@gmail.com

M. Rathinasamy et al. (eds.), *Water Resources and Environmental Engineering I*,
https://doi.org/10.1007/978-981-13-2044-6_15

satisfactory arrangements ensured by way of proper designing and planning of drainage system to ensure sustainability of road infrastructure.

The present study is though conducted on a stretch of NH-1, but it can be generalized and used for other roads having similar roadway and environmental conditions. In this study the size/capacity of the existing drain and the effect of increased contact time and traffic on bituminous mix in water are studied. In India, Bureau of Indian Standard [2] prescribes the test for stripping of coarse aggregate. The test is designed to capture the amount of stripping due to the presence of water on road surface.

2 Data Collection of Road Drainage

The road data collected from the field is given below:

Surface type	Bituminous
Shoulder Type	Paved/Bituminous and Earthen Shoulder
Carriageway Width of the road	12 m
Width of soil covered turf	1.3 m
Camber/cross slope on surface	0.025
Type of drain	Rectangular, cement concrete
Size of drain	1.0 m × 1.0 m
Average slope of drain bed	1 in 675
Length of road side drain	345 m.

3 Design of Road Side Storm Water Drain

For the design of road side drains first from the collected data we can calculate the time of concentration and critical rainfall intensity by the equation given by IRC: SP: 42 [3].

Time of concentration,

$$t_c = t_1 + t_2 + t_3, \tag{1}$$

where t_1 = time of flow of water on the road surface in the transverse direction in hours, t_2 = time of flow of water in the kerb channel in longitudinal direction up to the point of inlet in hours, t_3 = time of flow of water in the drain in longitudinal direction up to the point of outlet into a cross drain in hours.

Time of concentration has been calculated in Table 1 using the empirical formula given below:

Table 1 Time of concentration

Section	Direction of flow	L (km)	H (m)	t_1 (h) from Eq. (1)	t_2 (h)	t_3 (h)	t_c (h)	Speed (m/s)
Road surface	Transverse	0.020	0.520	0.013	–	–	–	0.42
Kerb channel	Longitudinal	0.030	0.1	–	0.040	–	–	0.21
Drain	Longitudinal	0.340	0.50	–	–	0.356	–	0.26
Time of concentration t_c (h) $t_1 + t_2 + t_3 =$							0.41 h 24.6 min	

Table 2 One hour maximum rainfall near Delhi

Frequency (T)	Rainfall in one hour (cm)	Conversion factor (F) (w.r.t. 2 year frequency)
2 years	3.6	1
5 years	5.5	1.53
10 years	6.2	1.72
25 years	8.0	2.22
50 years	9.2	2.56

Table 3 Conversion factors for rainfall intensity other than one hour

Duration of rainfall (min)	5	10	15	20	30	40	50	60	90
Rainfall intensity	13.32	10.26	8.64	7.49	6.01	4.79	4.21	3.6	3.0
Conversion factor	3.7	2.85	2.4	2.08	1.67	1.33	1.17	1.0	0.83

$$t_c = (0.81 \times L3/H)0.385, \tag{2}$$

where L = distance of critical point to the drain, H fall in level from the critical point to the drain level in meters.

For the given case of NH-1 the value of L is 340 m, which gives a concentration time of about 24.6 min, however a concentration time of 30 min seems to be more appropriate and therefore design is further checked upon 30 min.

One hour maximum rainfall near Delhi from rainfall maps of India according to IRC: SP: 42 [3] are given in Table 2.

Conversion factor for converting one hour rainfall intensity to intensity of other durations for 2 year frequency storm according to IRC: SP: 42 [3] are given below in the Table 3.

3.1 Determination of Design Discharge

The peak runoff is given by the following expression given by IRC: SP: 42 [3].

Discharge, Q = 0.028 PAI_c cum/s
Coefficient of Runoff, P = 0.9 (IRC: SP: 42)
Area of catchment, A = 12 × 340/10,000
= 0.40 ha

For t_c 30 min, critical intensity of rainfall from Table 3,

I_c = 6.01 cm/h
Therefore, Q (for 2 years frequency) = 0.028 × 0.9 × 0.4 × 6.01
= 0.06 cum/s
Q for 25 year frequency Table 1 = 0.06 × 2.22
= 0.134 cum/s

Adding 25% for discharge from intermediate drains, design discharge,

Q for 2 years frequency = 0.06 × 1.25
= 0.075 cum/s
Q for 25 years frequency = 0.134 × 1.25
= 0.168 cum/s.

3.2 Drain Section Calculations

Assume drain of 100 cm (width) × 55 cm (depth)

Free board = 15 cm
Depth of flow = 55 − 15
= 40 cm
Area of cross section of flow, A = 1.0 × 0.4
= 0.4 m^2
Wetted perimeter, P = 1.0 + 0.4 + 0.4
= 1.8 m
Hydraulic mean radius, R = A/P
= 0.222 m
Gradient of drain bed, S = 1 in 675

Manning's 'n' value for concrete drains above 600 mm diameter = 00.013

As per Manning equation, mean velocity, V = (1/n) ($R$2/3) ($S$1//2) = 0.94 m/s
which is OK as per IRC: SP: 42
Discharge, Q = A × V = 0.4 × 0.94
= 0.376 cum/s
which is more than 0.168 cum/s.

Therefore, drain size of 1.0 m × 1.0 m would be sufficient for 25 years frequency discharge.

Actual drain size provides is 1.0 m × 1.5 m, which is OK for 25 years frequency discharge. But it was found that water still stands out on the road surface.

4 General Drainage Problems on NH-1

Following problems were found to be causing standing off water on the road:

1. It has been observed that maintenance of drains is a neglected phenomenon.
2. The drains were found to be uncovered on many locations as shown in Fig. 1.
3. The covering slabs for some drains were found lying in drains as shown in Fig. 2.
4. Due to uncovered drains, silting, solid waste and garbage is found in the drains as shown in Fig. 3.
5. The entry point of drain not found made properly as shown in Fig. 4.
6. The slopes provided on the road were found to be inadequate to drain off water from road surface to drains which is clear from Fig. 5.
7. Stagnation of water takes place during rainy season as shown in Fig. 6.

Fig. 1 Uncovered drain

Fig. 2 Covering slab lying in the drain

Fig. 3 Dumping of garbage into drains

Fig. 4 Improper entry point of drains

Fig. 5 Adverse slope resulting into stagnation of water on the road

Fig. 6 Stagnation of water during rainy seasons

5 Stripping of Bituminous Mix

Due to improper drain maintenance and stagnation of water on the road it was found that stripping of aggregates from bituminous mix was taking place on the road. Details from concerned authorities were taken regarding type of aggregate and bitumen used on the road. Stripping tests were conducted as per IS: 6241:1971 [2] in the laboratory using coarse aggregates from Yamuna Nagar quarry and bitumen grade VG 30. Stripping of aggregate was tested by increasing the water immersion time from 1 to 5 days in order to determine the effect of extended contact time of aggregate with water on the value of stripping. The effect of traffic load and tyre friction was simulated by applying external pressure manually with a small tyre. The results are shown in Table 4.

Table 4 Stripping of aggregate due to extended contact time

S. No.	Contact time (Days)	Striping of aggregate (%)	
		Without pressure	With pressure
1	1	0	1
2	2	1	1
3	3	1	2
4	4	2	3
5	5	2.5	4

It is observed that the stripping value of aggregate increases with increase in immersion time of aggregate in water. Similarly, the stripping value increases with application of external pressure also.

6 Conclusions

The following conclusions are drawn from the study:

1. The time of concentration for the design of drain is 24.6 min and the design has been done for a concentration time of 30 min.
2. The critical intensity of rainfall for determination of design discharge is 6.01 cm/h.
3. It is observed that the present drain size of 1.0 m $\times$ 1.0 m is adequate for road drainage.
4. It is also observed that maintenance of drains was not proper. The problems found include uncovered and broken drains, stagnation of water, adverse cross slopes and silting of drains.
5. Due to stagnation of water stripping of aggregates was found to be taking place on the road and when tested in the laboratory the stripping value is found to be increasing with the increase in contact time of bituminous mix with water.
6. On the studied stretch of NH-1 stripping was observed whereas in the laboratory under standard conditions of stripping test no stripping was observed. However, when Standard conditions were varied to simulate the field conditions of longer inundation and effect of traffic, the stripping was found to increase substantially in laboratory also.

References

1. Rokade, S., Agarwal, P.K., Shrivastava, R.: Study on drainage related performance of flexible highway pavements. Int. J. Adv. Eng. Technol. (IJAET) **3**(1), 334–337 (2012)
2. Indian Standard (IS: 6241-1971): Method of Test for Determination of Stripping Value of Road Aggregate. Laboratory test, 1–4 (1971)
3. Indian Road Congress (IRC: SP: 42-2014): Guidelines of Road Drainage, New Delhi (2014)

MATLAB Code for Linking Genetic Algorithm and EPANET for Reliability Based Optimal Design of a Water Distribution Network

S. Chandramouli

Abstract Many researchers have developed different approaches for optimal design of water supply pipe networks. But, none of them provide a detailed coding for design procedure involved. Students and young researchers who are working in field of water distribution networks generally spend their valuable time searching for the procedure to link up Genetic Algorithm (GA) and EPANET. Therefore, in order to facilitate the young researchers and students, a detailed design procedure using EPANET solver with Genetic Algorithms in the MATLAB for reliability-based optimal design of water supply pipe networks is developed and presented in this paper with a case study.

Keywords MatLab · Epanet · Water distribution network · Reliability
Optimal design

1 Introduction

Optimal design of a pipe network is essential for any water distribution system since the 70% of the total cost is due to pipe network of the system. The practical importance and inherent complexities involved in the optimization of networks for distributing drinking water has attracted the attention of many in the past 30 years [1]. Basically, the optimization is the process of identifying a solution that is best in fulfilling the objective while satisfying the constraints. The objective may be minimizing cost or maximizing benefits. Numerous solution methods developed in the recent past for water pipe network optimization problem using different formulations such as Linear Programming [2–4] Nonlinear Programming [5, 6] heuristics and metaheuristic methods like Genetic Algorithm [7–11] Simulated annealing [12–14] Tabu search [15, 16] Differential evolution [17, 18] Memetic

---

S. Chandramouli (✉)
Department of Civil Engineering, MVGR College of Engineering (Autonoums), Vizianagaram, Andhra Pradesh, India
e-mail: chandramoulis@mvgrce.edu.in

M. Rathinasamy et al. (eds.), *Water Resources and Environmental Engineering I*,
https://doi.org/10.1007/978-981-13-2044-6_16

algorithm [19, 20] Cross-entropy [21] Scatter search [22] Immune algorithm [23] Shuffled frog leaping algorithm [24] Ant colony optimization [25–27] Particle swarm optimization [28] Harmony search [29–31] Particle swarm harmony search [32] etc.

GA is one of the search techniques widely used for optimization [8]. GA and Adaptive Cluster Covering with Local Search (ACCOL) algorithms showed their efficiency and effectiveness for optimal design of pipe networks. For large networks it is sensible to use a suite of algorithms in order to have the choice between fast-running algorithms and algorithms oriented towards more exhaustive but longer search. Abebe and Solomatine [33] and Wu and Simpson [34] applied one of the competent genetic-evolutionary algorithms—a messy genetic algorithm to enhance the efficiency of an optimization procedure. Prasad et al. [35], Prasad and Park [36] outlined a multi objective GA for the optimal design of a water distribution network to obtain the pareto-optimal front. Network resilience was used as a reliability measure to provide surplus head above the minimum allowable head at nodes. Tolson et al. [37] presented a new approach which links a genetic algorithm as the optimization tool with the first-order reliability method for estimating network capacity reliability and conducted a study to improve the efficiency of GA operational optimization through a hybrid method which combines the GA method with a hill climber search strategy. Two hill climber strategies, the Hooke and Jeeves and Fibonacci methods, were investigated. The hybrid method proved to be superior to the pure GA in finding a good solution quickly, both when applied to a test problem and to a large existing water distribution. Van Vuuren et al. [38] described the potential use of Genetic Algorithms for finding optimal diameters of a water distribution network.

A GA-based method for the least cost design of looped pipe networks for various levels of redundancy was presented by Nagesh Kumar et al. [39]. Konaka et al. [40] presented an overview and tutorial describing Genetic Algorithms developed specifically for problems with multiple objectives. Babyyan et al. [41] have focused to solve the problem of robust least-cost design of water distribution systems under uncertainty in input parameters by keeping the objective as to minimize the total design cost subject to a target level of system robustness. Vamvakeridou-Lyroudia et al. [42] used GA with fuzzy membership function for floating-on-the-system tank simulation for the optimization of pipe networks [43]. Presented a new approach in determining a penalty value depending on the degree of failure and the importance of the link supplying a specific node. Prasad [44] established a parameter for assessing the variations in demand satisfaction at different nodes of a drinking pipe network and developed a model for optimal and reliable design of water supply system suited for Indian conditions using GA. Afshar [45] proposed a compact GA to reduce the storage and computational requirements of population-based genetic algorithms.

It is clear from the above that many researchers have developed different approaches for optimal design of water supply pipe networks in particular with GA. But none of them provided a detailed design procedure involved. Therefore, in order to facilitate the young researchers and students, a detailed design procedure using EPANET solver with GA in MATLAB for reliability-based optimal design of a water supply pipe network. EPANET tool kit functions are best suited for analysis of a pipe network and one can easily work with these functions even though it has some limitations. MATLAB is best suited for all mathematical operations and also has the capability of linking external libraries. So in the present study, Matlab, GA tool kit functions are combined with EPANET tool kit functions. A detailed description of this is presented below.

2 Objective Function and Constraints

$$\text{Minimize } C_T$$

$$\text{Subjected to } (\text{TH})_i \geq (\text{TH})_{\min} \quad i = 1, 2, 3, \ldots, N \tag{1}$$

$$V_{\min} \leq V_j \leq V_{\max} \quad j = 1, 2, 3, \ldots, M \tag{2}$$

$$\text{NRP} \geq \text{NRP}_{\min} \tag{3}$$

$$\sum Q_k = 0 \quad k = 1, 2, 3, \ldots, P \tag{4}$$

$$\sum h_x = 0 \quad x = 1, 2, 3, \ldots, R \tag{5}$$

$$D_j \geq 0 \tag{6}$$

where N = Number of demand nodes in the network, M = Number of links in the network, P = Number of links joined at the node, R = Number of links in a loop, C_T = total cost of the network, TH = Total Head at the demand node in m, $\text{TH}_{\min}$ = Minimum HGL required at the demand node, NRP = Network Reliability Parameter, $V_{\min}$ = Minimum Velocity of flow in the link in m/s (0.6 m/s), $V_{\max}$ = Maximum Velocity of flow in the link in m/s (3.0 m/s), V = Actual Velocity of flow in the link in m/s, $\text{NRP}_{\min}$ = minimum Network Reliability Parameter.

3 EPANET Tool Kit

EPANET tool kit is a shared library which is a collection of functions. On Windows systems, the library is precompiled into a dynamic link library (.dll) file named EPANET2.dll. At run-time, the library is loaded into memory and made accessible to all applications. The Matlab Interface to Generic DLLs enables to interact with functions in dynamic link libraries. This interface has the ability to load an external library into Matlab memory space and then access any of the functions defined therein.

To load and unload EPANET library into MATLAB, the following functions are used.

loadlibrary('epanet2','epanet2')
unloadlibrary('epanet2')

Generally, at the beginning of the program, the library is to be loaded into memory and at the end of the program, it is to be unloaded. To invoke library functions, the calllib function is used. For loading input file and creating a report file, the following function is used.

calllib('epanet2','ENopen','input2.inp','report2.rpt',' ')
To assign a value to a link in the network, the following function is used.
calllib('epanet2','ENsetlinkvalue',1,0,100)
To analyze the network, the following functions are used.
calllib('epanet2','ENsolveH')
calllib('epanet2','ENsolveQ')
calllib('epanet2','ENreport')
To extract any value from the node/link, the following functions are used.
calllib('epanet2','ENgetnodevalue',1,11,0)
calllib('epanet2','ENgetlinkvalue',1,9,0)

4 GA Tool Kit Functions in Matlab

In MATLAB, the commands 'ga' and 'gatool' are used to implement GA to minimize an objective function. Ga implements the genetic algorithm at the command line to minimize an objective function where as 'gatool' opens a graphical user interface (GUI). 'ga' is used in the present study which is described below.

Syntax of 'ga':
x=ga(fitnessfcn, nvars, options)
'fitnessfcn' represents Fitness function
'nvars' indicate the number of independent variables in the fitness function
'options' represent options structure created with the gaoptimset, for example
options=gaoptimset;
gaoptimset('PopulationType','DoubleVector','PopulationSize',4,'InitialPopulation',[50;60;70;80],'PopInitrange',[50;100],'Generations',50)
ga(@fitnessfcn,4)
[x, fval, reason, output, population, scores]=ga(...) returns
Output—the output structure contains the following fields:
Population—returns matrix population, whose rows are the final population.
Scores—the scores of the final population.

5 Methodology

1. A standard benchmark network from the literature is selected.
2. An input file according to the format specified in the EPANET tool kit is prepared. A sample of the input file is given below.

[Title]
Analysis of a water supply pipe network
[Junctions]

ID	Elevation	demand	pattern
	1	150	1668

[Reservoirs]

ID	Head	Pattern
	1	210

[PIPES]

;ID	Node1	Node2	Length	Diam.	Roughness	Mloss	Status
1	1	2	1000	165.0	130	0	open

[OPTIONS]
UNITS LPM
HEADLOSS H-W
[END]

3. The EPANET.DLL, EPANET2.h and the input file are kept at the same location in which the Matlab files are stored in the computer.
4. The EPANET Tool kit functions are combined with the objective function through an M-file with the same name of that of the objective function. A sample file has been presented below.

```
function f=myfun(x,arg1,.....)
(function definition - if one needs to send arguments to
function, it can be incorporated in the function after x by
putting ','. Here 'myfun' is the function name, x is a vector of
decision variables and arg1 is argument)
```

path(path,‘c:\MATLAB7\extern\examples\shrlib’)
(Define the path of the external library stored in the computer.)
loadlibrary(‘epanet2’,‘epanet2’);
calllib(‘epanet2’,‘ENopen’,‘input2.inp’,‘report2.rpt’,‘’);
(loading EPANET tool kit)

d1=x(1); d2 = x(2);......
(assigning the diameters of the links generated by the ga to new variables)
calllib('epanet2','ENsetlinkvalue',1,0,d1);......
(Setting the diameters to the network)
calllib('epanet2','ENsolveH');
calllib('epanet2','ENsolveQ');
calllib('epanet2','ENreport');
(analyzing the network)
[a h2]=calllib('epanet2','ENgetnodevalue',1,11,0);......
[a v1]=calllib('epanet2','ENgetlinkvalue',1,9,0);......
(Extracting the information from the network after analysis)
ifge(h2,30)
arg1=0;
else
arg1=(30-h2);
end ......
if v1 >=3
arg2=(v1-3);
else
arg2=0;
end......
(Applying the conditions to be satisfied after the analysis)
calllib('epanet2','ENclose');
(Closing EPANET tool kit)
f=(632.7*x(1)^1.327+506.16*x(2)^1.327+759.24*x(3)^1.327+253.08*x(4)^1.327)
+1000*(arg1+arg2+......)
(Execution of objective function based on decision variables and the imposed constraints.)
end (End of the program)

5. The objective function developed in the M-file mentioned in step 4 is linked with Genetic Algorithm function through another M-file for which any name can be assigned.

To execute the objective function developed in the previous M-file, another M-file is to be developed with the following code.

```
options=gaoptimset;
gaoptimset
('PopulationType','DoubleVector','PopulationSize',4,'InitialPopulation',
[50;60;70;80],'PopInitrange',[50;100],'Generations',50)
ga(@myfun,4)
```

This M-file should be run to get the desired results.

6. The M-file created in Step 5 should be run to get the desired output.

6 Case Study

In the present study, a two-loop gravity network which is shown in Fig. 1 has been taken from literature to apply the present approach. It was first used by [2] for optimal design using LP. Subsequently the same network has been used by several researchers. Some of them are [3, 4, 8, 14, 24, 33, 44–48].

This network consists of eight links, six demand nodes, and one reservoir. Node 1 is a source node with HGL of 210 m and a demand of −1120 m^3/h. All the links in the network have a length of 1000 m and Hazen-Williams coefficient (CHW) is taken as 130. The minimum required HGL values and the costs of the pipes are taken from the literature [2].

7 Reliability Index

In the present study, the trapezoidal membership function is used to represent the satisfaction levels versus residual pressures available at the nodes and then reliability index at the node depending on the residual pressure beyond the minimum pressure requirement is determined. The residual pressures beyond the minimum pressures are categorized into three groups. The minimum and maximum residual pressures that may be obtained at anode are 0 and 30 m since the elevation difference between the first node and the last node is 30 m. The range of residual pressures and the corresponding satisfaction indices are shown in Table 1.

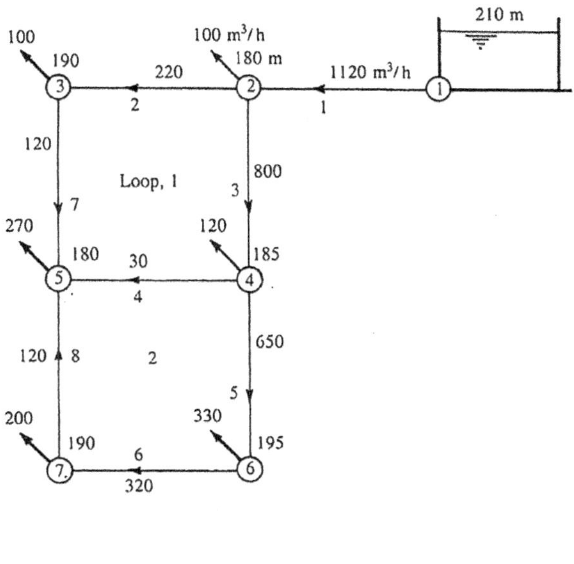

Fig. 1 Two loop pipe network (under Gravity)

Table 1 Range of excess residual pressures and their Satisfaction Indices

S. No.	Range of excess residual pressures (in m)	Reliability index	Remarks
1	0–15	$x/15$	x denotes the residual pressure beyond minimum
2	15–25	1	
3	25–30	$(30 - x)/5$	

8 Results and Discussions

The present method is applied on a two loop gravity network. Several optimal solutions are obtained. The details of the best seven optimal solutions are presented in the Table 2.

Total Cost (TC), Network Reliability Parameter (NRP) is used as performance indicators in the present study. The results obtained are presented in Table 2.

The optimal solutions obtained are fully satisfying the constraints of velocity and NRP except sol. 6 and sol. 7. As it is observed from TC and NRP that the NRP is decreasing with the decrease in cost. So depending upon the level of reliability, the cost of network depends.

9 Conclusions

A detailed methodology on linking of EPANET toolkit functions with GA tool box functions is presented in this paper. The present method is applied on a standard benchmark network (i.e., two-loop gravity network).

It is found that the method is easy and flexible. For every run of the program, one optimal solution is obtained. The run-time is less as well the function evaluation of number is also less. The use can change the code to suit his requirements. One can develop a pareto-optimal front by having more number of optimal solutions with corresponding reliability indictors. This methodology can be applied to any network for getting optimum pipe sizes subjected to minimum cost and maximum reliability.

Table 2 Results of the best solutions for the network

Parameter	Sol. 1	Sol. 2	Sol. 3	Sol. 4	Sol. 5	Sol. 6	Sol. 7
D1 in mm	508	508	457.2	457.2	457.2	508.00	457.2
D2 in mm	304.8	355.6	355.6	355.6	355.6	254.00	304.8
D3 in mm	406.4	406.4	406.4	406.4	406.4	406.40	406.4
D4 in mm	254	76.2	76.2	254	254	50.80	203.2
D5 in mm	355.6	355.6	355.6	355.6	355.6	355.60	406.4
D6 in mm	254	50.8	50.8	254	254	304.80	203.2
D7 in mm	254	406.4	406.4	304.8	304.8	254.00	254
D8 in mm	254	355.6	355.6	304.8	254	101.60	203.2
H2 in m	55.96	55.96	53.24	53.24	53.24	55.96	53.24
H3 in m	41.74	39.64	36.93	40.18	40.28	32.74	38.38
H4 in m	45.65	48.63	45.91	43.96	43.87	46.24	43.26
H5 in m	46.91	47.36	44.65	46.5	46.78	35.76	42.42
H6 in m	31.79	36.2	33.49	30.8	30.42	30.05	31.01
H7 in m	35.50	36.44	33.73	35.38	34.67	32.63	30.03
Th2 in m	205.96	205.96	203.4	203.24	203.24	205.96	203.24
Th3 in m	201.74	199.64	196.93	200.18	200.28	192.74	198.38
Th4 in m	200.65	203.63	200.91	198.96	198.87	201.24	198.26
Th5 in m	196.91	197.36	194.65	196.5	196.78	185.76	192.42
Th6 in m	196.79	201.2	198.49	195.8	195.42	195.76	196.01
Th7 in m	195.50	196.44	193.73	195.38	194.67	192.63	190.03
V1 in m/s	1.54	1.54	1.9	1.9	1.9	1.54	1.90
V2 in m/s	1.14	1.56	1.56	1.06	1.04	1.88	1.23
V3 in m/s	1.54	0.99	0.99	1.38	1.39	1.45	1.49
V4 in m/s	0.95	0.59	0.59	0.76	0.7	0.74	1.05
V5 in m/s	1.20	0.93	0.93	1.07	1.13	1.54	0.97
V6 in m/s	0.53	0.39	0.39	0.29	0.4	0.84	1.06
V7 in m/s	1.09	0.98	0.98	1.06	1.03	1.33	1.22
V8 in m/s	0.56	0.55	0.55	0.56	0.7	0.74	0.65
Q1 in lpm	18,670	18,670	18,670	18,670	18,670	18,670	18,670
Q2 in lpm	4985	9293	9293	6296	6176	5717	5384
Q3 in lpm	12,017	7708	7709	10,706	10,826	11,285	11,617
Q4 in lpm	2891	161	161	2309	2114	90	2046
Q5 in lpm	7126	5548	5548	6397	6712	9195	7571
Q6 in lpm	1626	48	48	897	1212	3695	2071
Q7 in lpm	3316	7625	7625	4628	4508	4049	3716
Q8 in lpm	1708	3286	3286	2437	2122	−361	1263
TC in units	498,000	543,000	503,000	504,000	486,000	450,000	461,000
NRP	0.591	0.6762	0.586	0.5705	0.5545	0.321	0.454

References

1. Reehuis, E.: Multiobjective Robust Optimization of Water Distribution Networks. A Master thesis submitted to Leiden Institute of Advanced Computer Science (LIACS), Leiden University, Niels Bohrweg 1, 2333 CA Leiden, The Netherlands (2010)
2. Alperovits, E., Shamir, U.: Design of optimal water distribution systems. Water Resour. Res. **13**, 885–900 (1997)
3. Quindry, G.E., Brill, E.D., Liebman, J.C.: Optimization of looped water distribution systems. J. Environ. Eng. Div. **107**, 665–679 (1981)
4. Kessler, A., Shamir, U.: Analysis of the linear programming gradient method for optimal design of water supply networks. Water Resour. Res. **25**, 1469–1480 (1989)
5. Shamir, U.: Optimal design and operation of water distribution systems. Water Resour. Res. **10**, 27–36 (1974)
6. Duan, N., Mays, L.W., Lansey, K.E.: Optimal reliability-based design of pumping and distribution systems. J. Hydraul. Eng. **116**, 249–268 (1990)
7. Dandy, G.C., Simpson, A.R., Murphy, L.J.: An improved genetic algorithm for pipe network optimization. Water Resour. Res. **32**, 449–458 (1996)
8. Savic, D.A., Walters, G.A.: Genetic algorithms for least cost design of water distribution networks. J. Water Resour. Plan. Manage. **123**, 67–77 (1997)
9. Gupta, A., Gupta, I., Khanna, P.: Genetic algorithm for optimization of water distribution systems. Environ. Model. Softw. **4**, 437–446 (1999)
10. Vairavamoorthy, K., Ali, M.: Optimal design of water distribution networks using genetic algorithm. J. Comput. Aided Civ. Infrastruct. Eng. **15**, 374–382 (2000)
11. Reca, J., Martinez, J., Banos, R., Gil, C.: Optimal design of gravity-fed looped water distribution networks considering the resilience index. J. Water Resour. Plan. Manage. **134**, 234–238 (2008)
12. Kirkpatrick, S., Gelatt, C.D., Vecchi, M.P.: Optimization by simulated annealing. Science **220**, 671–680 (1983)
13. Loganathan, G.V., Greene, J.J., Ahn, TJ.: Design heuristic for globally minimum cost water-distribution systems. J. Water Resour. Plan. Manage. **121**, 182–192 (1995)
14. Cunha, M.D.C., Sousa, J.: Water distribution network design optimization: simulated annealing approach. J. Water Resour. Plan. Manage. **125**, 215–221 (1999)
15. Glover, F.: Tabu search – Part 1. ORSA, J. Comput. **1**, 190–206 (1989)
16. Cunha, M., Ribeiro, L.: Tabu search algorithms for water network optimization. Eur. J. Oper. Res. **157**, 746–758 (2004)
17. Storn, R., Price, K.: Differential evolution—a simple and efficient Heuristic for global optimization over continuous spaces. J. Glob. Optim. **11**, 341–359 (1997)
18. Vasan, A., Simonovic, S.: Optimization of water distribution network design using differential evolution. J. Water Resour. Plan. Manage. **136**, 279–287 (2010)
19. Banos, R., Gil, C., Agulleiro, J.I., Reca, J.: A memetic algorithm for water distribution network design. Soft Comput. Ind. Appl. **39**, 279–289 (2007)
20. Banos, R., Gil, C., Reca, J., Montoya, F.G.: A memetic algorithm applied to the design of water distribution networks. Appl. Soft Comput. **10**, 261–266 (2010)
21. Perelman, L., Ostfeld, A.: An adaptive heuristic cross entropy algorithm for optimal design of water distribution systems. Eng. Optim. **39**(4), 413–428 (2007). https://doi.org/10.1080/03052150601154671
22. Lin, M.D., Liu, Y.H., Liu, G.F., Chu, C.W.: Scatter search heuristic for least cost design of water distribution networks. Eng. Optim. **39**, 855–876 (2007)
23. Chu, C., Lin, M., Liu, G., Sung, Y.: Application of immune algorithms on solving minimum-cost problem of water distribution network. J. Math. Comput. Model. **48**, 1888–1900 (2008)
24. Eusuff, M.M., Lansey, K.E.: Optimization of water distribution network design using the shuffled frog leaping algorithm. J. Water Resour. Plan. Manage. **129**, 210–225 (2003)

25. Maier, H., Simpson, A., Zecchin, A., Foong, W., Phang, K., Seah, H., Tan, C.: Ant colony optimization for design of water distribution systems. J. Water Resour. Plan. Manage. **129**, 200–209 (2003)
26. Zecchin, A.C., Simpson, A.R., Maier, H.R., Nixon, J.B.: Parametric study for an ant algorithm applied to water distribution system optimization. IEEE Trans. Evol. Comput. **9**, 175–179 (2005)
27. Zecchin, A.C., Simpson, A.R., Maier, H.R., Leonard, M., Roberts, A.J., Berrisford, MJ.: Application of two ant colony optimization algorithms to water distribution system optimization. Math. Comput. Model. **44**, 451–468 (2006)
28. Izquierdo, M., Izquierdo, J., Perez, R., Tung, M.M.: Particle swarm optimization applied to the design of water supply systems. Comput. Math. Appl. **56**, 769–776 (2008)
29. Geem, Z.: Optimal cost design of water distribution networks using harmony search. Eng. Optim. **38**, 259–277 (2006)
30. Geem, Z.W., Kim, H.H., Jeong, S.H.: Cost efficient and practical design of water supply network using harmony search. Afr. J. Agric. Res. **6**, 3110–3116 (2011)
31. Geem, Z.W., Cho, Y.H.: Optimal design of water distribution networks using parameter-setting-free harmony search for two major parameters. J. Water Resour. Plan. Manage. **137**, 377–380 (2010)
32. Geem, Z.W.: Harmony search optimization to the pump-induced water distribution network design. Civ. Eng. Environ. Syst. **26**, 211–221 (2009)
33. Abebe, A.J., Solomatine, D.P.: Application of global optimization to the design of pipe networks. In: 3rd International Conferences on Hydroinformatics, Copenhagen, Denmark, pp. 989–996 (1998)
34. Wu, Z.Y., Simpson, A.R.: Competent genetic evolutionary optimization of water distribution systems. J. Comput. Civ. Eng. **15**, 89–101 (2001)
35. Prasad, T.D., Sung-Hoon, H., Namsik, P.: Reliability based design of water distribution networks using multiobjective genetic algorithms. KSCE J. Civ. Eng. **7**, 351–361 (2003)
36. Prasad, T.D., Park, N.S.: Multiobjective genetic algorithms for design of water distribution networks. J. Water Resour. Plan. Manage. **130**, 73–82 (2004)
37. Tolson, B.A., Maier, H.R., Simpson, A.R., Lence, B.J.: Genetic algorithms for reliability-based optimization of water distribution systems. J. Water Resour. Plan. Manage. **130**, 63–72 (2004)
38. Van Vuuren, S.J., Van Rooyen, P.G., Van Zyl, J.E., Van Dijk, M.: Application and Conceptual Development of Genetic Algorithms for Optimization in the Water Industry. WRC Report No. 1388/1/05. Water Research Commission, Pretoria, South Africa (2005)
39. Nagesh Kumar, D., Raju, K.S., Ashok, B.: Optimal reservoir operation for irrigation of multiple crops using genetic algorithms. J. Irrig. Drain. Eng. **132**, 123–129 (2006)
40. Konaka, A., Coitb, D.W., Smithc, A.E.: Multi-objective optimization using genetic algorithms: a tutorial. Reliab. Eng. Syst. Saf. **91**, 992–1007 (2006)
41. Babyyan, A.V., Savic, D.A., Walters, G.A., Kapelan, Z.S.: Robust least cost design of water distribution networks using redundancy and integration based methodologies. J. Water Resour. Plan. Manage. ASCE, **133**(1), 67–77 (2007)
42. Vamvakeridou-Lyroudia, L.S., Savic, D.A., Walters, G.A.: Fuzzy Hierarchical decision support system for water distribution network optimization. J. Civ. Eng. Environ. Syst. **23**(3), 237–261 (2007)
43. Van Dijk, M., Van Vuuren, S.J., Van Zyl, J.E.: Optimizing water distribution systems using a weighted penalty in a genetic algorithm. J. Water SA (online) **34**(5) (2008)
44. Prasad, G.V.K.S.: Optimal design of water distribution network for uniform supply in intermittent system. A Ph.D. thesis submitted to NIT, Warangal, India (2008)

45. Afshar, M.H.: Application of a Compact Genetic Algorithm to pipe network optimization problems, SceitiaIranica. Trans. A: Civ. Eng. **16**, 264–271 (2009)
46. Fujiwara, O., Khang, D.B.: A two-phase decomposition method for optimal design of looped water distribution networks. Water Resour. Res. **26**, 539–549 (1990)
47. Liong, S.Y., Atiquzzaman, M.: Optimal design of water distribution network using shuffled complex evolution. J. Inst. Eng. **44**, 93–107 (2004)
48. Van Dijk, M., Van Vuuren, S., Van Zyl, J.: Optimising water distribution systems using a weighted penalty in a genetic algorithm. Water SA **34**, 537–548 (2008)

Regime-Wise Genetic Programming Model for Improved Streamflow Forecasting

K. Bhavita, D. Swathi, J. Manideep, D. Sree Sandeep and Maheswaran Rathinasamy

Abstract Forecasting of stream flow plays a vital role in flood forecasting studies, design, and operation of reservoirs. Several approaches such as physical models, conceptual models and statistical/black-box models are used to model complex uncertain peak flows in rivers. In the past, Genetic Programming (GP) have been a widely used for different hydrological applications. In this study we propose a regime-wise genetic programming model for efficient forecasting of streamflow during peak flows. In this approach, we first classify the flows into three regimes such as low, med and high based on their flow magnitude and develop separate GP models. The proposed approach was applied to a case study from Godavari River Basin, India. The results obtained show that the proposed approach of separate models for high flows performs better than the single model for all regimes.

Keywords Genetic programming · Streamflow forecasting · Flow regime

1 Introduction

Streamflow forecasting can be defined as prediction of flow rate in a stream in advance and also involves prediction of future inflow or flow rate of a particular stream [1]. Forecasting of flows has been an active area of research owing to its great importance in issuing flood warnings and providing the grounds for preventive actions. Generally stream flow predictions are based on observation of rainfall on the upper catchment, often supplemented by rainfall in the intervening catchments.

There are a variety of available methods for forecasting stream flows, which may fall into the following categories: process-driven methods and data-driven methods. Process–driven methods are simply a stream flow process in the view of a system theory, and mathematically approximate the internal physical process of the

K. Bhavita · D. Swathi · J. Manideep · D. Sree Sandeep · M. Rathinasamy (✉)
Department of Civil Engineering, MVGR College of Engineering, Vizianagaram
Andhra Pradesh, India
e-mail: maheswaran27@yahoo.co.in

M. Rathinasamy et al. (eds.), *Water Resources and Environmental Engineering I*,
https://doi.org/10.1007/978-981-13-2044-6_17

watershed system, that govern the stream flow process based on the physical process. On the other hand, data-driven methods are fundamentally black-box methods, without considering the internal physical mechanism, it generates a relationship among the input and output variables [1, 2]. The main disadvantage of the physics based models is that they are data intensive and computationally heavy. In many situations, the catchment under investigation may not be equipped with such data. In such scenarios, the data drive models have always been attractive alternative. It has the advantage of representing arbitrarily complex problems based on mathematical criteria by the use of the historical streamflow and rainfall values. These models are easy to apply for different conditions because the modeling and forecasting procedure is usually analogous. Owing to these impertinent advantages data-driven modeling techniques gain more and more popularity in the field of hydrology.

Some of the common data drive methods include Box Jenkins time series models [3], Artificial Neural Network models [4], Genetic Programming Models [1, 2], Support Vector Machine based models. Out of these different models, GP models are becoming more popular owing to their ability to develop a mathematical relationship between the predictor and the predicted variables. Generally, the historical flow values with certain lag is considered as the predictor variables and used for developing the model. While selecting the data for training, a single model is developed for all flow conditions. However, it is to be noted that the predictor–predict and relationship will be function of the flow condition and will be different during the high, medium and low flow conditions. In this paper an attempt is made to investigate efficiency of the regime-wise GP models where in separate models are developed for the different flow regimes. The proposed approach is tested on the observed streamflow data from Godavari River, India.

The rest of the paper is arranged in the following manner. Section 2 describes the brief outline of the GP approach and Sect. 3 provided information on the proposed approach. The study area is detailed in the Sect. 4. The results are discussed in Sect. 5.

2 Genetic Programming Model

Genetic programming is one of the recent computational techniques which is based on the process of natural evolution [5, 6]. The main principle involved in Genetic Programming (GP) is evolutionary algorithm where in the solution is arrived based on natural evolution theory which includes crossover, mutation. In GP, the candidate solutions are symbolic formulae which are used to stimulate the physical phenomena.

GP is a domain-free method that genetically yields a population of computer programs to solve a given engineering problem. The population of the programs are transformed into new generation of programs using genetic operations such as cross over and mutation. Unlike other regression methods, GP can automatically used to develop a model structure that fits the training data. A set of robust training data for learning, objective function, and functional and terminal set forms the input for the developing a GP models.

The depth of the functional set determines the complexity of the model and it includes the arithmetic operations such as, + , −, *, ^, /… and Boolean operations and conditionals. Further, the terminal set contains the arguments for the functions and can consist of numerical constants, logical constants, variables and others. The functional and terminals are chosen at random and constructed together to form a computer model in a tree like structure with a root node and branches extending from each function and ending in a terminal [5].

Initially, a random population is created and then the fitness of each individual is estimated. From these individuals, strong parents are selected and made to yield off spring following the process of cross over and mutation. The process of the generating offspring will continue iteratively till the desired goal in terms of the objective function is reached. An example of crossover and mutation is given in Figs. 1 and 2 respectively. The detailed procedure involved in the implementation of GP is given in [5].

It is possible in GP approach to look through all of the best programs and analyze how many times each input variables appears in a way that contributes to the fitness of the program that contain them.

3 Methodology

At first the entire data set sub-divided into three flow regimes such as high, medium, and low flow based on their discharge values. Flow values <4 m^3/s is considered as low; flow values between 4 and 90 m^3/s is taken as medium flows whereas values >90 m^3/s is considered to be high flows. For each of the regimes, individual GP based models were developed and the results were compared with the single GP model using the following performance measures.

(1) Coefficient of Determination

$$R^2 = 1 - \frac{\sum_{t=1}^{N} (\text{Actual}(i) - \text{Forecast}(i))^2}{\sum_{t=1}^{N} (\text{Actual}(i) - \overline{\text{Actual}})^2}$$

(2) Root Mean Square Error (RMSE)

$$\text{RMS} = \sqrt{\frac{1}{n}\left(\sum_{i=1}^{n} x_i^2\right)}$$

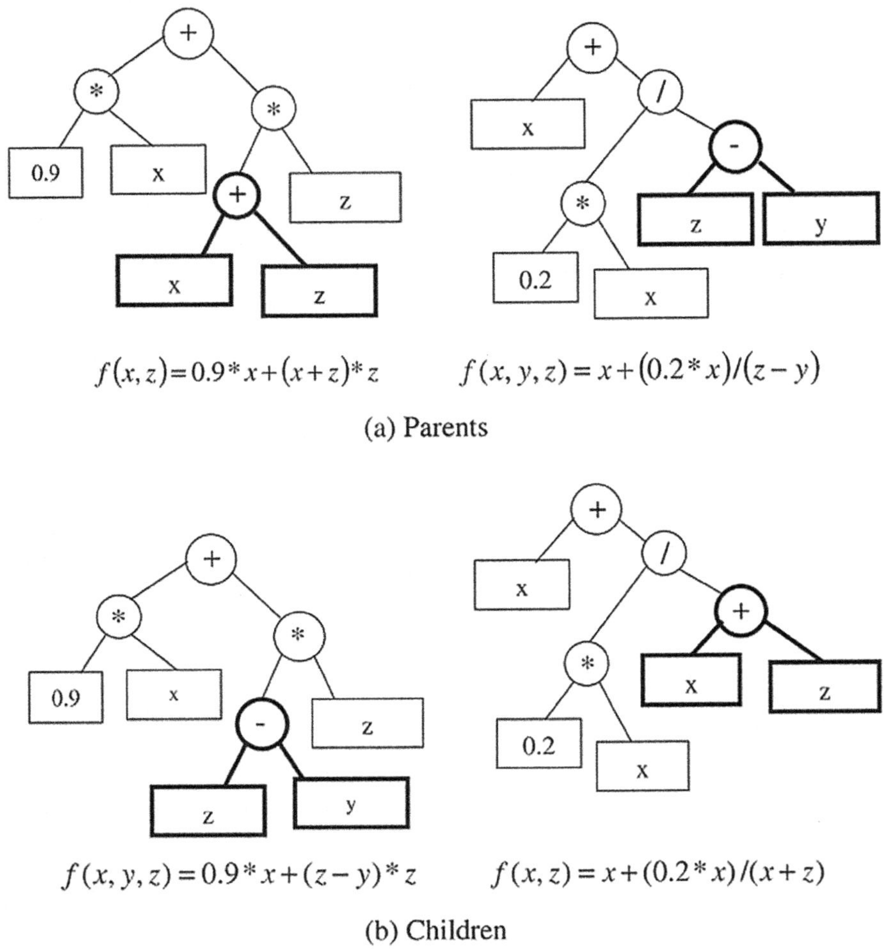

Fig. 1 Cross over operation

(3) Normalized Root Mean Square Error (NRMSE)

$$\text{NRMSE} = \frac{\text{RMSE}}{Q_{\text{mean}}} = \frac{\text{RMSE}}{\left(Q_{\text{high}} - Q_{\text{low}}\right)}$$

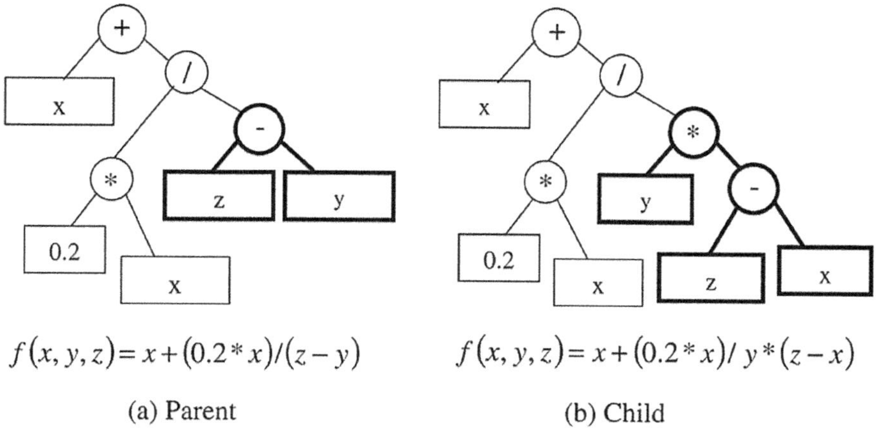

Fig. 2 Mutation operation

4 Study Area

In this study, forecasting was developed for Godavari river basin focusing on the Asthi station. The data for the implementation of models was obtained from India-WRIS (http://www.india-wris.nrsc.gov.in/ website). The total available data was found out to be 47 years and out of this 50% is used for training, 25% for testing and remaining for validation purposes. For evaluating GP programs, we have used Discipulus tool (Algorithm tool). The function set of the system can be composed of arithmetic operations (+, −, /, ∗), and function calls (f ∈ {ex, x, sin, cos, tan, log, sqrt, ln, power}. Each function implicitly includes an assignment to a variable v[i], which facilitates the use of multiple program outputs in LGP, whereas in tree-based GP, this needs to be incorporated explicitly. However in this study we did not use the complex function like conditionals, data transfer etc.

In this work, we have considered the historical discharge values of particular stream which is needed to be forecasted as input data. For the input data we have considered the previous discharge values for almost 47 years for forecasting daily and monthly lead. In order to achieve the accurate predictability the input data is assigned in such a way that 60% of discharge data is allocated under training data, 25% of discharge data is allocated under testing data and remaining 25% of discharge data is allocated under validation data. Based on these data the GP model runs and gives the output in the form of a program, using which the model for forecasting is developed. Using the above procedure case studies are done on the concept of stream flow forecasting using GP on Godavari River basin.

Table 1 Performance statistics for GP models

Model	Coefficient of distribution (R^2)	Root mean square error (RMSE) (m^3/s)	Normalized root mean square error (NRMSE) (m^3/s)
Model-I			
Complete data	0.758	6.84	2.4255×10^{-4}
Model-II			
High flow regime	0.744	29.91	6.0941
Medium flow regime	0.912	0.061	1.318×10^{-3}
Low flow regime	0.899	0.0046	1.659×10^{-7}

5 Results

In this study, the basic equation used for developing model for univariate stream flow forecasting is given below

$$Q_{t+1} = f_n(Q_t, Q_{t-1}, Q_{t-2}\cdots\cdots)$$

The one day forecast is estimated as a function of the previous day's flows. The following table is the results (in terms of R^2) of the models developed.

The Model-I denotes the GP model developed for the complete data, where as in Model-II, GP models were developed for like high, medium and low flows individually. The result statistics for all the models are shown in Table 1. It can be seen that the individual models have better performance that the Model-I. Further to investigate the performance of the Model-I in capturing the peak flow, the results were separated and the result statistics was estimated and reported in Table 2. It can be observed that the individual model for high flow performed better than the Model-I predicted results.

Table 2 Comparison of high flow prediction using complete data set and individual models

Model	Coefficient of distribution (R^2)	Root mean square error (RMSE) (m^3/s)	Normalized root mean square error (NRMSE) (m^3/s)
High flows predicted using Model-I	0.70	37.78	8.078
Model-II-High flow regimes	0.744	29.91	6.0941

6 Conclusion

The main aim of the study is to propose a regime-wise Genetic Programming model for improving the prediction accuracy of the high flows. The proposed approach was applied to forecast discharge data at Asthi station on Godavari River. The results show that GP predicted the stream flow with quite good agreement with the observed data. It was observed the regime-wise model performed better than the model developed using the complete data. As further recommendations, one can work out how the regime-wise model perform at higher lead times.

References

1. Rabunal, J.R., Puertas, J., Rivero, D.: Determination of the unit hydrograph of a typical urban basin using genetic programming and artificial neural networks. Hydrol. Process. **21**(4), 476–485 (2006)
2. Selle, B., Mutil, N.: Testing the structure of a hydrological model using genetic programming. J. Hydrol. **397**(1–2), 1–9 (2010)
3. Parasuraman. K.A.: Toward improving the reliability of hydrologic prediction: model structure uncertainty and its qualification using ensemble-based genetic programming framework. Water Resour. Res. **44**(12) (2008)
4. Jyothiprakash, V., Magar, R.B.: Multi-step ahead daily and hourly intermittent reservoir inflow prediction using artificial intelligence techniques using lumped and disturbed data. J. Hydrol. **450**, 293–307 (2012)
5. Babovic, V., Keijzer, M.: Rainfall runoff modelling based on genetic programming. Nord. Hydrol. **33**(5), 331–346 (2002)
6. Koza, J.R.: Genetic programming as a means for programming computers by natural selection. Stat. Comput. **4**(2), 87–112 (1994)

Author Index

M. Rathinasamy et al. (eds.), *Water Resources and Environmental Engineering I*,
https://doi.org/10.1007/978-981-13-2044-6

Printed by Printforce, the Netherlands